Organized in the framework of FJNSP LIA and EFES

French-Japanese Symposium on Nuclear Structure Problems

Organized in the framework of FJNSP LIA and EFES

French-Japanese Symposium on Nuclear Structure Problems

RIKEN, Wako, Japan, 5–8 January 2011

Edited by

Hideaki Otsu
RIKEN Nishina Center, Japan

Tohru Motobayashi
RIKEN Nishina Center, Japan

Patricia Roussel-Chomaz
Grand Accélérateur National d'Ions Lourds, France

Takaharu Otsuka
University of Tokyo, Japan

World Scientific

NEW JERSEY • LONDON • SINGAPORE • BEIJING • SHANGHAI • HONG KONG • TAIPEI • CHENNAI

Published by

World Scientific Publishing Co. Pte. Ltd.
5 Toh Tuck Link, Singapore 596224
USA office: 27 Warren Street, Suite 401-402, Hackensack, NJ 07601
UK office: 57 Shelton Street, Covent Garden, London WC2H 9HE

British Library Cataloguing-in-Publication Data
A catalogue record for this book is available from the British Library.

NUCLEAR STRUCTURE PROBLEMS
Proceedings of the French-Japanese Symposium

ISBN 978-981-4417-94-5

Printed in Singapore by Mainland Press Pte Ltd.

PREFACE

The French-Japanese Symposium on Nuclear Structure Problems was held at RIKEN Nishina Center from 5th to 8th, January 2011. It was organized in the framework of FJNSP LIA (the French-Japanese International Associated Laboratory for Nuclear Structure Problems) and EFES (the JSPS core-to-core project "Exotic Femto Systems").

The aim of this symposium was to discuss on the topics listed below and to enhance and initiate collaborations between Japan and France in the relevant field.

- Structure of exotic nuclei
- Nuclear reactions of stable and unstable nuclei
- Nuclear astrophysics
- Superheavy elements
- New facilities and equipment
- On-going and planned collaboration
- Related topics

100 participants attended the symposium, with 67 participants from Japan, 25 participants from France and 8 from other countries. 66 talks in total were given, and very active and fruitful discussions were made during the whole symposium with interactions between Japanese and French communities.

Collaboration between France and Japan in nuclear physics has a long history. Various collaborative studies have been conducted among researchers and institutions in France and Japan. The new scheme of collaboration LIA has been established in January 2008. The EFES project has supported several joint meetings and schools. Rich results from collaborations were mentioned by several speakers. Series of joint symposia have been held since 1970s, and the latest one was at the Institut Henri Poincaré in Paris in fall 2008, the 150th anniversary of the governmental relationship between Japan and France. That was the first symposium since the

LIA collaboration started. The present one is the second symposium. In this proceedings, we could record recent results including the ones by the collaborations between the two countries.

The symposium was also sponsored by Center for Nuclear Study (CNS), University of Tokyo and RIKEN Nishina Center. We finally wish to express our thanks to the conference secretaries, T. Iwanami and Y. Kishi for their continuous and invaluable works.

January 2012,

H. Otsu (RIKEN)
T. Motobayashi (RIKEN)
P. Roussel-Chomaz (CEA Sacley)
T. Otsuka (CNS, University of Tokyo)

CONTENTS

GANIL-SPIRAL2: A NEW ERA

SYDNEY GALES
GANIL, DSM-CEA/IN2P3-CNRS, Bd. Henri Becquerel, F-14076 Caen cedex, France

GANIL presently offers unique opportunities in nuclear physics and many other fields that arise from not only the provision of low-energy stable beams, fragmentation beams and re-accelerated radioactive species, but also from the availability of a wide range of state-of-the-art spectrometers and instrumentation. An overview of the physics with secondary beams carried out at GANIL is presented. Selected examples of recent experiments using fragmentation of high energy intense stable heavy ions beams and re-accelerated SPIRAL1 "exotic" beams and the associated instruments are used to illustrate the ongoing physics program. With the construction of SPIRAL2 over the next few years, GANIL is in a good position to retain its world-leading capability. As selected by the ESFRI committee, the next generation of ISOL facility in Europe is represented by the SPIRAL2 project to be built at GANIL (Caen, France). The future prospects of the accelerator complex GANIL-SPIRAL1 and the path towards SPIRAL2 is also briefly introduced.

1. Introduction

In this paper, I would like to present the recent highlights of the physics of "exotic" nuclei and/or the main results obtained recently on nuclei far off stability at GANIL, a emerging field which we as, physicists we can compare to the exploration of a new territory. The perspectives offer by the SPIRAL2 project at GANIL will be shortly introduced in the last paragraph of this paper.

Since its beginning in 1983, GANIL [1] has been involved in active research in the field of exotic nuclei. At GANIL the available stable ions beams range from ^{12,13}C to Kr isotopes with intensities up to 2pμA for ^{13}C at 95MeV/n and 0.1pμA for Kr at 50 MeV/n (beam power between 3 to 6 KW on target). For heavier ions (Lead and Uranium) an energy range between 5-25 MeV/n with 10^{10}pps can be achieved. Light and medium mass ion beams raised to high energy are fragmented on a thin target, giving rise through the "fragmentation process" to light exotic nuclei.

The successful program of the physics of "exotic" nuclei produced by the so-called "in-flight " method was extended recently towards new possibilities offered by high-quality, low-energy Radioactive Ions Beams (RIB) at the SPIRAL facility [1] using the "ISOL" production method.

This SPIRAL facility, which became operational in September 2001, can be summarized as follows: the high energy, high power beams extracted beams from the three GANIL accelerators C01 (2)-CSS1-CSS2 is sent to a carbon target. The incident nuclei are fragmented by nuclear reactions in this target, generating a population of radioactive nuclei which diffuse out of this hot target (2 000°C). After ionization in an ion source, the particles are injected into the CIME cyclotron to be accelerated to energies of 1.7 to 25 MeV per nucleon. Since its start-up, various ions have been produced and accelerated in the SPIRAL facility, ranging from exotic nuclei as light as ^{8}He to heavier nuclei like ^{76}Kr. About 200 shifts per year are devoted to experiments with SPIRAL beams.

At GANIL, the rather unique capabilities, using both "In flight" and "ISOL" techniques, to produce a large variety of light and medium "exotic" species have boosted the scientific program.

2. Exotic nuclei at GANIL: recent physics highlights

As intensities of secondary Radioactive Ion Beams are usually four or five orders of magnitude lower than typical intensities of stable beams, physics with secondary RIB requires very specific detection devices. Reactions and Structure studies of "exotic" nuclei are carried out using intensively the high energies of the fragmented "in flight" RIB from GANIL, the post accelerated ISOL-SPIRAL1 RIB and low energy (5-10 MeV/n) intense Uranium beams. The beams are coupled to spectrometers like LISE, VAMOS and/or 4π–γ array called EXOGAM, charged particle detectors like MUST and TIARA. Those combinations are ideal tools for the studies of nuclear structure and reactions.

In the following we will present a few recent highlights of the physics results achieved in the last few years.

2.1. *Tunnelling of exotic systems*

A modern variation of the Rutherford experiment to probe the tunnelling of exotic nuclear matter from the measurement of the residues formed in the bombardment of ^{197}Au by extremely neutron-rich ^{8}He nuclei was carried out. Using a novel off-beam technique the most precise and accurate measurements of fusion and neutron transfer involving reaccelerated unstable beams are reported. The results show unusual behaviour of the tunnelling of ^{8}He compared to that for lighter helium isotopes, highlighting the role of the intrinsic structure of composite many-body quantum systems and pairing correlations [2].

2.2. *Another major research area at GANIL is the quest for shell structure changes far off stability*

Experiments on shell structure evolution around magic number 8, 20 and 28 have been recently investigated using both “in flight” and ISOL RIB via knock out or transfer reactions associated to gamma spectroscopy.

For the N=14 shell, measurements were performed using 10.5 MeV/n ^{24}Ne beam (10^5 pps) on a 1 mg cm^{-2} CD_2 target .The charged particle array TIARA was used to detect protons, in combination with four segmented clover detectors of the EXOGAM array placed at 90° around the target. The heavy transfer residues ^{25}Ne were detected and identified in the focal plane of VAMOS. The high efficiency of the whole system allowed three-fold coincidences data (g +proton+ ^{25}Ne) to be acquired. Excitations energies, l transfer and spectroscopic factors were measured for states up to 4.0 MeV. The salient features of this work can be summarized as follows [3]:

The N=14 gap is large in ^{25}Ne, the N=20 gap formed between the $d_{3/2}$ and fp intruder states seems to be small and the N=28 gap has greatly increased in the Ne isotopes. This lead to an inversion between the $f_{7/2}$ and $p_{3/2}$ orbit, as inferred from the energies of the 3/3- and 7/2- states.

For the N=28 neutron shell, the energies of the excited states in very neutron-rich ^{42}Si and $^{41\text{-}43}P$ have been measured using in-beam γ-ray spectroscopy from the fragmentation of secondary beams of $^{42\text{-}44}S$ at 39 MeV/n. The low 2^+ energy of ^{42}Si, 770(19) keV, together with the level schemes of $^{41\text{-}43}P$ provide evidence for the disappearance of the Z=14 and N=28 spherical shell closures, which is ascribed mainly to the action of proton-neutron tensor forces [4]. Recently the evolution of the structure of N=28 nuclei from the doubly-magic ^{48}Ca nuclei to the strongly deformed ^{42}Si has been investigated at GANIL. Of particular interest is the low lying level scheme of nuclei with Z<20 and N=28 like for example ^{44}S. In the neighbouring nuclei N=29, the spectroscopy study of ^{43}S have led to the discovery of a low lying 7/2- isomeric state, very weakly connected to the deformed 3/2- ground state demonstrating a shape coexistence at low excitation energy in ^{43}S [5]. The N=28 isotone ^{44}S have been recently investigated at GANIL.

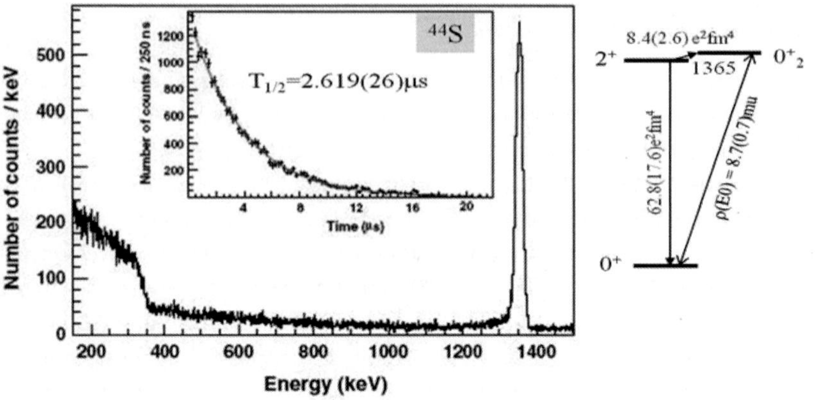

Figure 1. Electron Energy Spectrum showing the 0^+_2—0^+_1 E0 transition in ^{44}S Left the low excitation energy spectrum of ^{44}S [6]

A low-lying second 0^+_2 isomer has been clearly identified in ^{44}S through the measurement of the 0^+_2 –0^+_1 transition as shown in figure.1 and its lifetime measured (2,62 μs). From the measured B (E2) reduced transition probability and the extracted Monopole strength ρ^2, it has been shown [6] that the two 0^+ state are rather pure, in agreement with the conclusion that ^{44}S a spherical prolate shape transition is evidenced. The spectroscopy of ^{43,44}S support the previous finding, namely a shell erosion of the N=28 gap in that region of mass.

2.3. *MUST2 Campaign at RIKEN*

Under the framework of the International Associate Laboratory (LIA), the MUST2 collaboration (GANIL, IPN Orsay and SPhN/IRFU/Saclay) and the RIKEN Nishina Center physics group have formed a new collaboration.

MUST2@RNC collaboration [7] prepared an experimental campaign using the unique opportunities offered by the "in flight" intense light "exotic" beams produced by the new facility RIKEN-RIBF and the Big-Rips fragment separator. Two direct reactions studies have been investigated using the reactions ^{11}Li (d, ^{3}He) ^{10}He and ^{24}O (p, p'). The light particles were detected by the charged particles array MUST2 consisting of a set of Si strips detectors placed around the gas target.

This campaign illustrates the power of coupling of state of the art detectors array, new and intense exotic beams to investigate the structure of nuclei even beyond the stability line.

2.4. *A new spin aligned pairing phase in 92Pd*

In nuclei with equal neutron and proton numbers (N=Z), enhanced correlations arise between neutrons and protons that occupy orbitals with the same quantum numbers. Such correlations have been predicted to favour an unusual type of nuclear super fluidity, termed isoscalar neutron–proton pairing, in addition to normal isovector pairing. Despite many experimental efforts, these predictions have not been confirmed.

An international collaboration have achieved an experimental observation of excited states in the N=Z=46 nucleus ^{92}Pd. Gamma-rays emitted following the ^{58}Ni (^{36}Ar, 2n) ^{92}Pd fusion–evaporation reaction were identified using a combination of state-of-the-art high-resolution γ-rays, charged-particle and neutron detector systems (EXOGAM,DIAMANT,NEDA). The results reveal evidence for a spin-aligned, isoscalar neutron–proton coupling scheme, different from the previous prediction (see Ref [8] and figure. 2). The authors suggest that this coupling scheme replaces normal super fluidity (characterized by seniority coupling) in the ground and low-lying excited states of the heaviest N=Z nuclei. Such strong, isoscalar neutron–proton correlations would have a considerable impact on the nuclear level structure and possibly influence the dynamics of rapid proton capture in stellar nucleosynthesis.

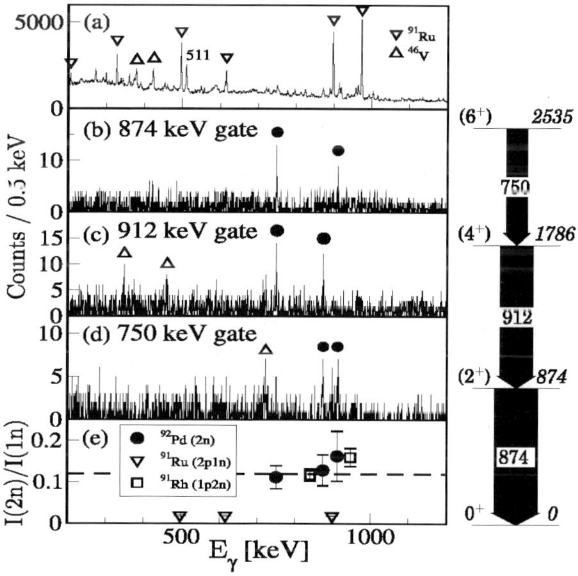

Figure 2. Gamma rays spectrum of the N=Z nuclei 92Pd at the basis for the evidence of a new np coupling scheme [8].

2.5. *Fission Time Measurements as New Probe for testing Super heavy Element Stability.*

The so-called "super heavy" elements (comprising more than 110 protons) are generally formed by fusion reactions between two lighter nuclei. One of the main difficulties of these attempts to synthesize Super Heavy Elements (SHE) resides in the excitation which is unavoidably generated within these nuclei, in the form of temperature and deformation, when they are formed by fusion. Fission events with times greater than 10^{-18} s (a billionth of a billionth of a second) were observed for nuclei with 120 and 124 protons. These compound systems were formed by bombarding nickel and germanium targets with uranium ions accelerated at the GANIL. They were identified by means of INDRA, a large array charged particle detector, which covers almost the entire space surrounding the targets. Using blocking techniques in the crystal-like targets of Ni and Ge and by comparing the yields of the cross-sections for quasi-elastic and fusion-fission events, very long half-life times were observed for these very large Z compound systems (~10^{-18}s). Although this 10^{-18} s time is clearly very short, on the scale of nuclear lifetimes, it is long enough to sign unambiguously the formation of elements having 120 and 124 protons, and to characterize them as being highly stable with respect to fission, when they are not excited [9,10]. More recently, the same compound Z=120 system was investigated by means of Ni+U reaction. During the collision and the formation of the compound system, electronic rearrangement of the system leads to X-ray emissions which are characteristics of the formed system. By detecting with INDRA the fission fragments in coincidence with X-rays, a clear evidence of an X-ray energy corresponding at Z=120 was observed [11]. The analysis is in progress. These results open up new perspectives in the race for super heavy elements, and the quest for the "island of stability".

2.6. *Prompt gamma spectroscopy of fully identified fission fragments with VAMOS*

Last but not least, thanks to the high intensity and low energy ^{238}U beams coupled to the VAMOS spectrometer and the gamma array EXOGAM, it has been possible for the first time to carry out a *prompt gamma spectroscopy of fully identified* fission *fragments.*

The reaction ^{238}U+ C at 6.1 MeV/n was carried out. Fission of U or Pu isotopes was induced in the collision. The fission fragments were detected and identified at the VAMOS focal plane by means of an ionization chamber (Z), time of flight, magnetic rigidity (M/Q) and total kinetic energy (M).

The focal plan detection of the VAMOS spectrometer was upgraded to achieve Z and A resolution for fission fragments both in A and Z as shown in figure 3.

The target was surrounded by a set of Si detectors to identify light inelastic or transfer fragments (SPIDER array) and a set of high efficiency Ge detectors from EXOGAM to achieve prompt spectroscopy of identified fission fragments.

Measurements of prompt Doppler-corrected de-excitation γ-rays from uniquely identified fragments formed in fusion-fission reactions of the type ^{12}C (^{238}U, ^{134}Xe) Ru were achieved [12].

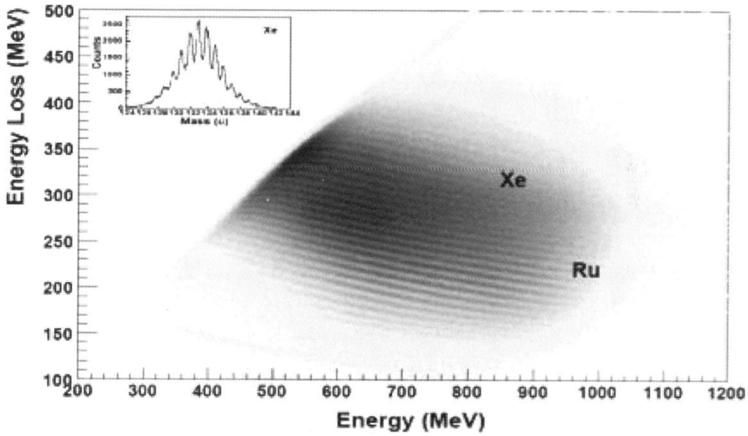

Figure 3 : Two dimensional spectrum of ΔE vs. E_{total} in the $^{238}U+^{12}C$ system at an incident beam energy of 1.45 GeV. The inset shows the associated mass spectrum for the Xe isotopes.

3. The future of GANIL – SPIRAL: SPIRAL2

Since the beginning of the SPIRAL project it was proposed to enlarge the range of RIB produced through the fission of uranium target and the delivery of high intensity neutron–rich fission fragments. This idea is now being concretized in the SPIRAL2 project [13].

The construction of SPIRAL 2 started in the middle of 2005 and is supported by the EU FP7 through the Preparatory Phase contract since 2008. The first beams are expected in the year 2013. As an added bonus, the facility will also produce over 10^{15} neutrons/s, making it the world's most powerful source of fast neutrons for several years.

A rather complete presentation of the current status of the project and of scientific case of SPIRAL2 facility will be presented in this meeting by M. Lewitowicz [13]).

Acknowledgments

I would like to thank my colleagues from the physics group at GANIL, for allowing me to use some of the results of their work in this paper.

References

1. www.ganil-spiral2.eu/.
2. A. Lemasson and al., Phys. Rev. Lett. 103, 232701 (2009).
3. W. Catford and al., Phys. Rev. Lett. 104, 192501 (2010).
4. B. Bastin and al., Phys. Rev. Lett. 101, 092501 (2008).
5. L. Gaudefroy and al., Phys. Rev. Lett. 102, 092501 (2009).
6. C. Force and al., Phys. Rev. Lett. 105, 102501 (2010).
7. D. Beaumel, private communication.
8. B. Cederwall and al., Nature,09644 (2010)
9. D. Jacquet and M. Morjean, Progress in Particle and Nuclear Physics 63, 155-185 (2009).
10. M. Morjean and al., Phys.Rev.Lett.101, 072701 (2008).
11. M. Morjean, private communication.
12. A. Shrivastava and al., Phys. Rev. C80, 051305 (2009).
13. www.ganil.fr/research/developments/spiral2/ and M. Lewitowicz, Proceedings of this workshop.

NUCLEAR PHYSICS IN JAPAN AND RIKEN NISHINA CENTER

Hideto En'yo
RIKEN Nishina Center for Accelerator-Based Science

Brief introduction to Nishina Center for Accelerator-Based Science and the present status of Japanese nuclear physic are presented.

1. Nishina Center in history and at present

In 1931, 80 years ago, Dr. Yoshio Nishina started his own laboratory in RIKEN and constructed the first cyclotron in Japan. This laboratory is the real root of the present Nishina Center. Japanese often celebrate 80th anniversary as "Parapluie Célébration" just because the Chinese character for umbrella looks like a combination of eight and ten. We wished to have a large celebration, but it became difficult due to the disaster happened in March 11th, 2011. Nevertheless this year must be very special for Nishina Center.

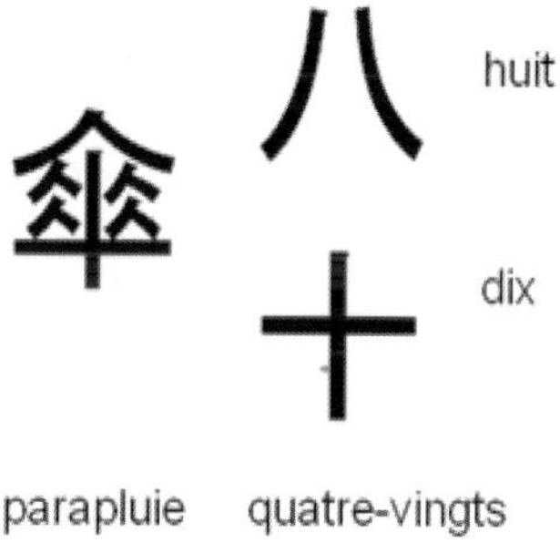

Figure 1: Chinese characters for umbrella, eight and ten.

Since the inauguration of Nishina Laboratory in 1931, RIKEN has build 9 cyclotrons. The first 4 are in history (see Fig.2), but 5th to 9th cyclotrons are alive and operational in the present RIKEN RIBF(Radio Isotope Beam factory). The first cyclotron was completed in 1937, and the second one in 1944. The second cyclotron was one of the largest in the world at the time, but abandoned into Tokyo Bay after the World War II. The third one was in operation in 1952, with most of the parts recuperated from the first one. The fourth cyclotron was in 1966 as the post-war recovery symbol of the accelerator-based science in Japan, presently resting in piece as a monument in the RIKEN Wako Campus.

The preset RIKEN RI Beam Factory (RIBF) consists of two linear accelerators and 5 cyclotrons. The 5th and 6th cyclotrons (AVF and RRC) with the LINAC (RILAC-I) form "old facility" became operation in 1990. In 2006 we have completed the 7th, 8th and 9th cyclotrons (FRC, IRC and SRC) in cascade operation with the old facility. This was the born of RIBF. Needless to say the heart of RIBF is SRC, Super conduction Ring Cyclotron of K2600-MeV which is the most powerful cyclotron in the world. Synchronizing with the start of SRC, we have inaugurated Nishina Center, by gathering all the accelerator-related researches in RIKEN. Now Nishina Center's activity includes not only RIBF but also our branches in Brookhaven National Laboratory in USA and Rutherford Appleton Laboratory in England.

Figure 2: The first to the fourth cyclotrons built in RKEN, from left top to right bottom. Dr. Nishina is sitting on the second cyclotron.

Since 2007 we are delivering beams for users while continuously building experimental facilities and tuning the accelerators for more intense and stable beam delivery. As any other accelerators in the world, the performance of RIBF increases year by year. By 2010 we are successful to deliver 1000pnA for lighter ions(deuteron to ^{18}O), 230pnA for ^{48}Ca, 30pnA for ^{83}Kr and 0.8pnA for Uranium(238). We plan to reach 10pnA for Uranium by 2012 with RILAC-II, which was successfully commissioned a few months ago, and many other improvements being underway. The design goal of 100pnA will be reached in 2014-2015.

2. Nuclear Physics in Japan

Talking about all the accelerators in Japan, there are four accelerators which are very prominent in the world. They are RIBF at Wako, J-PARC in Tokai, Belle in Tsukuba and SPRing-8 in Harima. The RIBF of RIKEN Nishina Center, now the most powerful heavy- ion accelerator in the world, chases physics of element genesis, i.e. the physics of unstable nuclei. J-PARC became operational very recently. Last autumn (2010), they run the first beam time for hadron physics experiment. It will have the highest proton beam power in a few years. They are also expecting many spallation-neutron users and muon users from life and material science. Their facility was damaged seriously by the earthquake, but we expect them to be back on line in January 2012. Belle stopped its operation in July 2010, after recording the highest luminosity of e^+e^- colliders and Super Belle is approved for construction. Spring-8 at Harima is one of the best photon factories in the world. There are many other good accelerators but these four are most valuable for the world. Eighty years of history since Nishina brought us to this very competitive situation.

All of these have different scientific specialties. But, having these top-rated machines (hence costly) domestically, all the owner institutes are forced to segregate themselves scientifically, i.e. to minimize overlaps in the scientific coverage with others and to maximize cooperation with others.

From nuclear physics community's point of view, we have 6 accelerator facilities available in Japan as illustrated in Fig. 3. We have 4 institutions holding the accelerators for the nuclear physics community. Besides RIKEN-RIBF and Tokai J-PARC, Research Center for Nuclear Physics in Osaka University is operating K=400 Ring Cyclotron. Tohoku University very recently changed the organization of their university-owned electron machines for the inter-community use. They named the new organization as *ELPH* (Research Center for *ELectron* and *PHoton* Science) characterized with the 1.2GeV electron stretcher. National Institute for Radiological Science is operating

800MeV heavy ion synchrotron. This machine is versatile and unique (designed carried from NUMATRON*). Major beam time is spent for medical use and limited beam time is available for nuclear physics. SPRing-8 (8 GeV electron Synchrotron) is certainly build for material and bio science. Presently there is only one beam line for nuclear physics, LEPS, built by Osaka University to deliver the laser- Compton photons, renown with the discovery of the penta-quark. Osaka University together with RIKEN Nishina Center is preparing the next beam line LEPS2 with the same technique but 10 times stronger intensity and with a new large spectrometer.

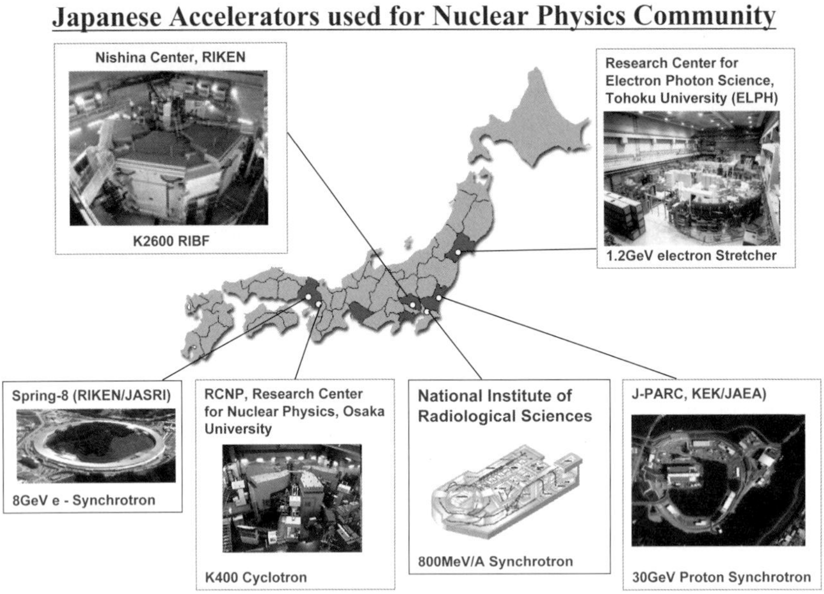

Figure 3: Accelerators in Japan available for nuclear physics community. RIBF, J-PARC, RCNP and ELPH(Tohoku-U) are build for nuclear/particle physics.

The mechanism to determine the long-range scientific plan in Japan is a bit unclear. The Science Council of Japan, created in 1949, is the prestigious bottom-up organization by scientists. The Council used be a driving force to promote big science in Japan and successful to inaugurate many inter-community type laboratories such as INS(Institute of Nuclear Study, University

* Nuclear physics project proposed in 1974 from Institute of Nuclear Study, the University of Tokyo, characterized with a double-ring synchrotron for GeV/A heavy ions. It was not approved for construction but its philosophy was carried over by many new programs in nuclear physics.

of Tokyo) and KEK. The Council had once lost its prestigious position in the governmental policy decision, because scientists themselves could not make pin-point recommendation any longer as the scale of scientific budget got bigger in many scientific fields. With these regret the Science Council of Japan recently modified itself to re-establish the mechanism of bottom-up policy making, and called all the scientific communities to prepare long-range plans which require a large-scale facility. The plans are reviewed in the Council and successful plans will be recommended to the government. Japanese Nuclear Physics Community has endorsed the four plans as listed in Table 1. The 1st and 2nd plan, the upgrade plans for RIBF and JPARC, are well represented in the Council review. The third plan from RCNP, "Ultra high sensitive measurement for the study of origin of matter dominated universe", and the fourth plan from RCNP and ELPH together, "Quark nuclear physics with photon beams" will be reviewed soon or later. The third one is a sort of declaration of RCNP to switch their direction a bit toward the fundamental physics, and the forth one is to create a syndicate of physics using electrons and photons. Those 4 plans more or less determine the direction of nuclear physics community for coming 10 years.

Table 1. Long range plans submitted to the Science Council of Japan from the Japanese nuclear physics community.

Title	Description	by
Towards "Island-of-Stability" via upgrades of the RIKEN Radioactive Isotope Beam Factory	RI intensity upgrade. ISOL.	RIKEN
Study of the origin of matter by upgrading J-PARC (Advancing Hadron physics at J-PARC by RIKEN)	JPARC Accelerator upgrade. Facility upgrade (extension of counting hall)	KEK, RIKEN, RCNP
Ultra high sensitive measurement for the study of origin of matter dominated universe	Intensity upgrade. Double beta decay. UCN-EDM. Muon-electron conversion.	RCNP
Quark Nuclear Physics with Photon Beams	Spring-8 new beam line. Two institutes cooperate to cover 0.6-3GeV photon	ELPH, RCNP

3. Summary

As described in this manuscript the future of scientific challenges in nuclear structure problems and other fundamental questions in nuclear physics will largely rely on RIKEN-RIBF and its future growth. Apparently world-wide collaboration is the key to the growth. We will soon (within a year) complete the new experimental facilities, SCRIT(Self Contained RI Ion Target) and SAMURAI (Superconducting Analyzer for MUltiparticles from RAdio Isotope beams), and intensity upgrade for Uranium (and other heavy species). In near future (within a few years) we will facilitate 1)Rare-RI Mass Ring to measure the mass precisely, 2) Slow-RI which slow down the produced RIs for beta NMR studies and other related subjects, and Return-Beam Transport to bring back high energy beam to the old facility building. Presently RIBF is the most powerful heavy ion accelerator in the world. To prepare other new RI beam facilities in the world become operational, RIBF major upgrade is being planned for 5-10 years of scale.

After the 3.11 earthquake, Japan is facing very difficult situation. We are in electric power shortage and the Tohoku recovery plan may cost us as a significant budget cut in the fundamental science. Nevertheless RIKEN Nishina Center will do our best to deliver beams for user as much as possible. Your cooperation is very much appreciated.

Nuclear Physics Programs at RIBF

H. Sakurai

RIKEN Nishina Center for Accelerator-Based Science
2-1 Hirosawa, Wako, Saitama, 351-0198, Japan
E-mail: sakurai@ribf.riken.jp

I report present status and perspectives of "RI Beam Factory (RIBF)" at RIKEN. RIBF is a top world-class in-flight RIB facility. Emphasis is given to on-going and future programs at the new facility through introducing experimental devices and recent highlights.

Keywords: Spectroscopy on exotic nuclei, Fast radioacitvie beam techniques

1. Introduction

RIBF is a top world-class RIB facility, which was constructed to establish a new framework of nuclear physics, to elucidate the origin of elements, and to explore new applications with fast RIBs.[1] To achieve these goals, the RIBF facility consists of a high-power accelerator complex, a large-acceptance in-flight separator and unique devices to cover various experimental programs.

As illustrated in Fig. 1, the accelerator complex consists of the old facility and a newly constructed one.[2] The old facility has a K540-MeV ring cyclotron (RRC) and two injectors, a variable-frequency heavy-ion linac (RILAC) and a K70-MeV AVF cyclotron. A new high-power heavy-ion booster system consisting of three ring cyclotrons with K=570 MeV (fixed frequency, fRC), 980 MeV (intermediate stage, IRC), and 2600 MeV (superconducting, SRC), respectively, boosts energies of the RRC beams up to 440A MeV for light ions and 350A MeV for very heavy ions. The goal of the primary beam intensity is set to be 1 pμA. At the end of 2006, the first primary beam was successfully accelerated at and extracted from SRC. The new in-flight separator BigRIPS[3] is to convert the SRC primary beams into RIBs. Combination of the high-power accelerator complex and the in-flight separator makes it possible to explore the limit of of nuclei and provides high yield rates of exotic nuclei.

The RIBF facility has started delivering RI beams since 2007. In 2007, the first commissioning with a 350A MeV ^{238}U beam successfully produced and identified two new isotopes, ^{125,126}Pd, even with a low intensity of 0.03 pnA.[4] In November 2008, the second attempt to search for new isotopes was performed by utilizing the U beam with a 10 times higher intensity of 0.3 pnA and 45 new isotopes were produced and identified for 4 days.[5] These results have demonstrated high performances of RIBF in RIB production and identification.

To maximize the potentials of intense RI beams available at RIBF, several new devices in the new facility are proposed.[6,7] A few devices have already started to provide physics outputs, as shown in Fig. 2. A multi-function beam line spectrometer (ZeroDegree) and a high resolution spectrometer (SHARAQ)[8] were launched in 2008 and 2009, respectively. Other experimental devices, SAMURAI,[9] SCRIT,[10] Rare-RI ring,[11] SLOWRI[12] and IRC-to-RIPS beam line,[13] as shown in Fig. 1, are under construction or to be funded. All of these devices give unique opportunities at RIBF in terms of energy ranging from several 10 keV to a few 100A MeV as well as of a wide coverage of scientific programs.

In this talk, recent highlights obtained at BigRIPS/ZeroDegree setup[3,14] and future plans are introduced and discussed.

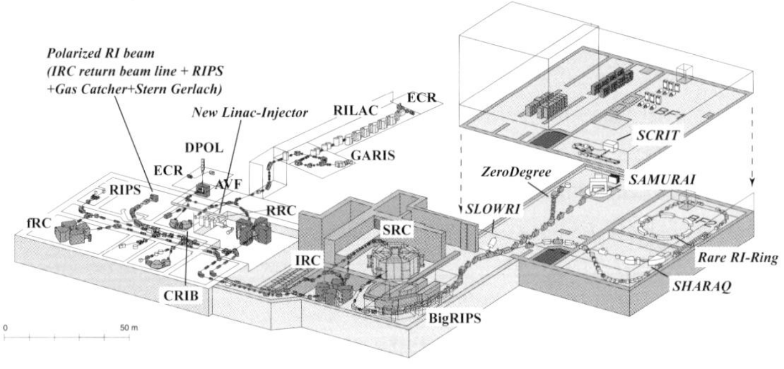

Fig. 1. A schematic view of the existing facility (left-hand side) and the RIBF under construction (right-hand side). The italic-labeled installations are major ones planned in the phase-II program of the RIBF project.

2. Reaction study with island-of-inversion nuclei and its vicinity

In December 2008, the first spectroscopic experiments for exotic nuclei in the island-of-inversion region were performed in a "DayOne" campaign by

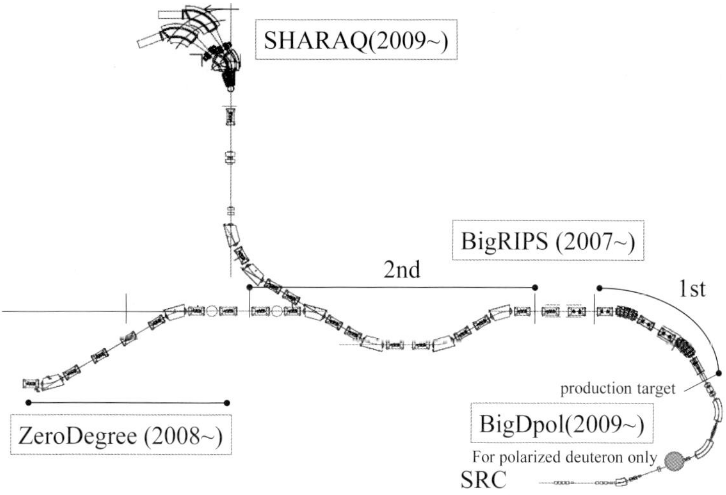

Fig. 2. Experimental devices available at the new facility of RIBF.

utilizing intense RIBs produced with a 345A MeV ^{48}Ca beam. The maximum ^{48}Ca beam intensity was as high as 170 pnA. Typical beam intensities for the exotic nuclei were, for example, 10 and 5 counts/s for ^{31}Ne and ^{32}Ne, respectively, for an intensity of 100 pnA. Compared with the yield rate of ^{31}Ne in the first identification experiment at RIPS,[15] that at BigRIPS is 10^5 times higher. This great gain for RIB intensities has demonstrated that the new facility of RIBF is indeed the next generation facility.

In the DayOne campaign, three experimental programs were performed, in-beam gamma spectroscopy by H. Scheit et al.[16,17] to determine the energy of low lying excited states in ^{32}Ne and other nuclei, inclusive Coulomb-breakup measurement by T. Nakamura et al.[18] to search for possible halo-nuclei in the island-of-inversion region and in a light mass region, and total interaction cross section measurement by Takechi and Otsubo et al.[19] to investigate exotic forms of nuclear matter in the neutron-rich Ne isotopes.

One of the highlights in these experiments suggests that the degree of deformation for the neutron-rich Ne isotopes increases as a function of neutron number.[16] Possible roles of weakly bound neutrons in collectivity enhancement would be very interesting subjects for future works at RIBF. In addition, discovery of the halo nucleus ^{31}Ne in the deformed region[18,19] gives a trigger to consider a possible mechanism of deformed halo states, which has been already discussed in theoretical works.[20]

Further studies for the exotic structure would be made possible via the intense ^{48}Ca beam. At the end of 2010, the three kinds of programs were carried out successfully to get exotic properties in more neutron-rich nuclei such as ^{38}Mg and ^{42}Si.

Other promising results associated with island-of-inversion are from proton-inelastic scattering on ^{24}O with the 'MUST2' array,[21] moment measurements of micro-second isomeric states,[22] and production of spin-aligned RI beam with secondary beams.[23] It should be noted that these works are organized under France-Japan collaborations.

3. Decay spectroscopy on $A\sim110$ neutron-rich nuclei

The first decay spectroscopy on fission fragments produced via the in-flight fission of ^{238}U beam was performed in December, 2009. The maximum primary beam intensity was as high as 0.8 pnA, and the intensity on average was 0.2-0.3 pnA. In this experiment proposed by Sumikama and Nishimura et al.,[24,25] half-life measurement, beta-gamma spectroscopy, and isomer spectroscopy were accomplished at the same time, where a decay-spectroscopy setup was mounted at the end of ZDS; four germanium clover detectors and two $LaBr_3$ detectors for gamma-rays, and nine stacks for double sided silicon strip detectors to detect beta-rays and heavy-ions.

One of recent results published is a half-life measurement associated with the r-process path, and 18 new data for neutron-rich nuclei with $A\sim110$ have been reported.[24] The other is spectroscopy on ^{109}Nb,[25] where combination of isomer-observations and beta-gamma coincidence measurements was employed to observe low-lying levels in ^{109}Nb.

4. Programs with Deuteron Beams

Specific programs with deuteron beams are polarized-deuteron and proton elastic scattering for three-body interactions[26] and production of deeply-bound pionic atoms via (d,^{3}He) reaction.[27]

The 3NF study via deuteron and proton elastic scattering with a 250A MeV polarized deuteron beam was performed in April, 2009.[28] In this experiment, the AVF-RRC-SRC acceleration scheme was applied to accelerate the polarized deuteron beams. A single turn extraction of the beam was successfully achieved, and the polarization amplitudes were as high as 80%. A polarimeter has been newly constructed to obtain a complete set of deuteron analyzing power in the scattering. The polarimeter was installed at the beam transport line between the exit of SRC and the production target at BigRIPS. Other data-sets at different and higher energies will be obtained in future.

In the (d,^{3}He) program, the entire BigRIPS will be utilized as a high resolution spectrometer employing the missing mass technique. The missing mass spectrometer mode was examined in May, 2009.[29] Further analysis would give future upgrade plans to improve optical conditions including the SRC and BT line.

5. Summary and Outlook

Special emphasis has been given to recent highlights in the programs for studying nuclear structure and reactions at several experimental devices. One of research highlights at RIBF is to obtain brand-new results in the island of inversion region and $A{\sim}110$ region by utilizing the intense ^{48}Ca beam and ^{238}U beam, respectively. The great gain of RIB intensities in that region has demonstrated that the new facility of RIBF is indeed the next generation facility.

Other promising primary beams, such as Ge, Kr and Xe, are to be used for programs approved at the BigRIPS/ZeroDegree. In 2011, a new linac RILAC-II coupled with the SC-ECRIS has been completed to deliver more intense heavy-ion beams. This new linac is utilized as an injector for Xe and U beams for SRC and the old linac RILAC could be dedicated for super heavy element programs at the GARIS spectrometer;[30] the element 113th production and decay spectroscopy on the super heavy nuclei. In addition to the primary beam upgrades, two of new experimental devices, the SAMURAI spectrometer[9] and the SCRIT system,[10] will be completed in FY2011, and are giving more and more physics opportunities at RIBF.

References

1. <http://www.rarf.riken.go.jp/Eng/>
2. Y. Yano, Nucl. Instrum. Methods B **261**, 1009 (2007).
3. T. Kubo, Nucl. Instrum. Meth. B **204**, 97 (2003); <http://www.nishina.riken.jp/UsersGuide/BigRIPS>
4. T. Onishi et al., J. Phys. Soc. Japan **77**, 083201 (2008).
5. T. Onishi et al., J. Phys. Soc. Japan **79**, 073201 (2010).
6. Technical information on the devices are found in <http://rarfaxp.riken.go.jp/RIBF-TAC05>
7. H. Sakurai, Eur. Phys. J. ST **150**, 249 (2007).

8. T. Uesaka, Contribution to this symposium; T. Uesaka et al., Nucl. Instrum. Methods B **266**, 4218 (2008).
9. K. Yoneda, Contribution to this symposium; Contribution to the 8th International Conference on Radioactive Nuclear Beams, Grand Rapids, USA, May, 2009.
10. T. Suda, Contribution to this symposium; M. Wakasugi et al., Nucl. Instrum. Methods A **532**, 216 (2004); M. Wakasugi et al., Phys. Rev. Lett. **100**, 164801 (2008); T. Suda et al., Phys. Rev. Lett. **102**, 102501 (2009).
11. Y. Yamaguchi et al., Nucl. Instrum. Methods B **266**, 4575 (2008).
12. M. Wada et al., Nucl. Instrum. Methods A **532**, 40 (2004).
13. H. Ueno, Eur. Phys. J. ST **162**, 221 (2008).
14. H. Sakurai, Nucl. Instrum. Methods B **266**, 4080 (2008).
15. H. Sakurai et al., Phys. Rev. C **54**, R2802-R2805 (1996).
16. P. Doornenbal, H. Scheit, N. Aoi et al., Phys. Rev. Lett. **103**, 032501 (2009).
17. P. Doornenbal, H. Scheit, N. Aoi et al., Phys. Rev.C **81**, 041305R (2010).
18. T. Nakamura et al., Phys. Rev. Lett. **103**, 262501 (2009).
19. M. Takechi et al., International Conference on Nucleus-Nucleus Collision, August, 2009, Beijing, China; M. Takechi et al., Nucl. Phys. A **834**, 412c (2010).
20. M. Misu et al., Nucl. Phys. A **614**, 44 (1997); I. Hamamoto, Phys. Rev. C **69**, 041306R (2004).
21. V. Lapoux, Contribution to this symposium.
22. J.M. Daugas, Contribution to this symposium.
23. H. Ueno, Contribution to this symposium.
24. S. Nishimura et al., Phys. Rev. Lett. **106**, 052502 (2011).
25. H. Watanabe et al., Phys. Lett. B **696**, 186–190 (2011).
26. K. Sekiguchi et al., Phys. Rev. Lett. **95**, 162301 (2005).
27. K. Suzuki et al., Phys. Rev. Lett. **92**, 072302 (2004).
28. K. Sekiguchi et al., Phys. Rev. C **83**, 061001 (2011).
29. K. Itagaki, private communication.
30. K. Morita, Mod. Phys. Lett. A 25, Sp. 1804 (2010).

UNBOUND STATES OF THE DRIP-LINE NUCLEUS ^{24}O

V. LAPOUX*, S. BOISSINOT, E. C. POLLACCO, F. FLAVIGNY, C. LOUCHART, L. NALPAS and A. OBERTELLI

CEA-Saclay, DSM/IRFU SPhN, Gif-sur-Yvette, F-91191, France
**E-mail: vlapoux@cea.fr*

H. OTSU, H. BABA, R.J. CHEN, N. FUKUDA, N. INABE, D. KAMEDA, M. MATSUSHITA, T. MOTOBAYASHI, T. ONISHI, E.Y. NIKOLSKII, M. NISHIMURA, H. SAKURAI, M. TAKECHI, S. TAKEUCHI, Y. TOGANO, K. YONEDA, A. YOSHIDA and K. YOSHIDA

RIKEN Nishina Center, 2-1 Hirosawa, Wako, Saitama 351-0198, Japan

A. MATTA, Y. BLUMENFELD, S. FRANCHOO, F. HAMMACHE, PH. ROSIER and E. RINDEL

IPN-Orsay, CNRS-IN2P3, Orsay, F-91406 France

P. GANGNANT, CH. HOUARNER, J.F. LIBIN and F. SAILLANT

GANIL, CEA/DSM - CNRS/IN2P3, Bld. H. Becquerel, Caen, F-14076, France

The characteristics of the new N=16 shell gap at the neutron drip-line can be deduced from the neutron excitations of ^{24}O. An experiment was carried out to investigate the unbound excited states of ^{24}O using the proton elastic and inelastic proton scattering. It was performed in the BigRIPS line and combines the unique intensities of the RIBF ^{24}O beam with the state-of-the-art particle detector array MUST2. The method is explained.

Keywords: Unstable nuclei; shell structure; magic number; inelastic scattering.

1. Resonances of the nuclei close to or at the drip-line

1.1. *Motivations*

The nuclear models disagree for the predictions of the spectroscopy of nuclei far away from the valley of stability, due to the different assumptions on nucleon-nucleon interactions and few-body correlations. The studies of the structure and spectroscopy of nuclei with a high ratio of N/Z, close to the drip-lines, are important for our understanding of the evolution of the nuclear shell structure with the isospin. Drip-line nuclei also combine several aspects like very diffuse nuclear surfaces, halos or neutron skins, resonant states, new shell gap effects. They are weakly-bound with few or no bound excited states, and the coupling to the continuum is playing a significant role since the scattering states are much closer to the continuum states than those in stable nuclei. All these aspects are stringent tests for the microscopic structure calculations, and the drip-line nuclei appear as benchmarks for the nuclear models [1]. The existence and position of the low-lying resonant states in the light exotic nuclei would provide crucial information to constrain and to test the validity of the models.

1.2. *Resonances of the light neutron-rich ^{6}He halo nucleus*

^{6}He has low neutron thresholds (S_n = 1.87 and S_{2n} =0.97 MeV) and no bound excited states. Under 12 MeV, only the characteristics of the first excited state, a 2^+ at 1.8 MeV($\Gamma = 113$ keV) have been firmly established. Various models predict a series of 2^+, 1^+, 0^+ states above the 2_1^+, but disagree on the energies of these states. We have investigated the spectroscopy of ^{6}He using the 2-neutron transfer ^{8}He(p,t)^{6}He reaction. The experiment was carried out at GANIL with the SPIRAL ^{8}He beam at 15.4 A.MeV, with no contaminant. The average intensity was 1.8×10^4 part/s. The reaction target was a 50 μm-thick foil of polypropylene $(CH_2)_n$. The beam profile and incident angle on the target were monitored event by event by two beam tracking detectors, CATS. The light charged particles were measured using 5 position-sensitive Si-strip telescopes MUST2 [2]. One MUST2 module was composed by a first stage of a 300μm double-sided Si-strips detector (DSSD) with an active area of 10*10 cm^2 and 128(x,y) strips, each strip providing energy and time-of-flight (TOF) measurements. The DSSD is backed up by a 4 cm-thick CsI-crystal detector, providing the remaining energy Er. Each strip detector has a minimum energy threshold of 0.5 MeV with intrinsic position, angular and energy resolutions of 0.76 mm, 0.3° (at 15 cm from the target) and 45(5) keV (FWHM), respectively. The particles can be identified using the standard correlation techniques between the energy deposited in the Si-strip stage, ΔE, and TOF, and between ΔE and Er [2]. The excitation energy (Ex) spectra were calculated by the missing mass method from the measured total kinetic energy and angle of the triton detected in coincidence with forward scattered particles, either ^{6}He or ^{4}He. The analysis is described in Ref. [3], two new resonances were observed: at 2.6 ± 0.3 MeV (width $\Gamma = 1.6 \pm 0.4$ MeV) and at 5.3 ± 0.3 MeV with

$\Gamma = 2 \pm 1$ MeV. The angular distributions were analyzed in the framework of the coupled-reaction-channel (CRC) calculations. This method was successful to describe consistently the previous data set measured at 15.6 A.MeV: ^{8}He(p,p), (p,d) and (p,t) to the ^{6}He g.s. and 2_1^+ [4]. It was found that the new states correspond to a 2^+ and to a $L = 1$ states, respectively. The resonances were compared with the results obtained recently by various theoretical frameworks, the ab-initio calculations [5] and the models including an explicit treatment of the continuum couplings (CC) of bound and scattering states [6–8]. These CC models give calculated positions of the states rather close to the experimental values but they overestimate the width. The new states, challenging the nuclear theories, could be used as benchmarks for checking the microscopic inputs of the newly improved structure models.

1.3. *Study of the shell effect associated to N=16*

Experimentally the neutron drip-line is known up to the oxygen chain, with ^{24}O being the last bound isotope. In the region of the neutron-rich nuclei around ^{24}O, the neutrons occupy the sd-fp shells and new effects have been found, indicated or predicted for the shell structure, in contrast with the shell occupation known for the stable nuclei. Experimentally, N=16 has been indicated as a new possible magic number [10], theoretically interpreted as an enhancement of the neutron shell gap between the $2s_{1/2}$ and $1d_{3/2}$ subshells, with respect to the sd shell in the stable nuclei, this effect being driven by the tensor proton-neutron force [11]. To explore the properties of this new shell effect, the low-lying spectroscopy of ^{24}O was studied: at GANIL, no gamma-ray associated to ^{24}O was found [12], showing that the excited states are unbound. At MSU, via invariant mass method from ^{24}O fragments and neutrons, possible states were indicated [13] around 4.5 and 5.3 MeV, above the neutron threshold S_n at 4.19 MeV [9], but not clearly identified. An alternative technique to study the neutron excitation is to measure the proton elastic and inelastic scattering, and to use the particle spectroscopy.

2. Spectroscopy of ^{24}O using the (p,p') reaction

From the detection of the recoil proton the (p,p') kinematics can be reconstructed and the Ex spectrum can be obtained via the missing mass method. Our ^{24}O(p,p') measurement became feasible with the advent of the RI Beam Factory operated by RIKEN Nishina Center and CNS, University of Tokyo. The experiment (RIBF57) was performed in 2010 using the MUST2 detector array and the secondary beams produced by the fragmentation on a Be target of the ^{48}Ca beam at 345A.MeV. The beams were separated using the BigRIPS line [14]. With the exceptional rates of ^{48}Ca, (60 up to 180 particle nA), ^{24}O was produced at 263A.MeV with unique intensities, 1100/s on average (2000, maximum).

2.1. *Experimental set-up and kinematics*

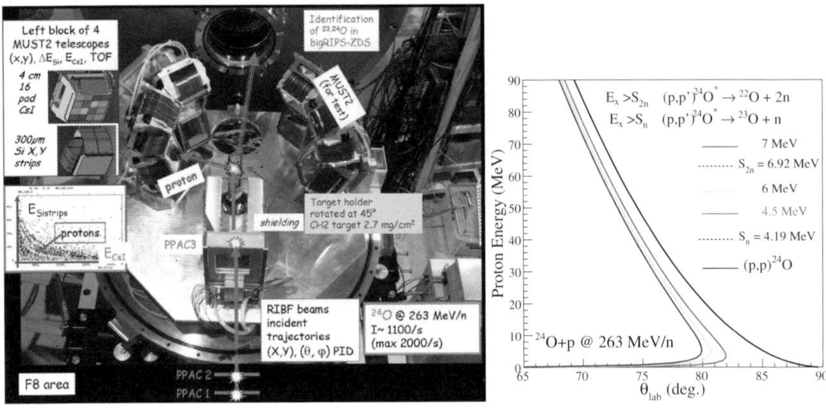

Fig. 1. Top view of the set-up inside the reaction chamber; plot of the ^{24}O (p,p') kinematics calculated for three hypothetical states, located at 4.5 MeV, 6 and 7 MeV.

The reaction chamber was mounted in the F8 area with 8 MUST2 detectors, assembled in two blocks. The 2.7 mg/cm^2-thick (CH_2) reaction target was rotated at 45 deg. towards the left block. The left telescopes were located at 23 cm downstream of the target, to cover lab. angles between (65-90) deg. The experimental set-up located in the F8 area is presented in Fig. 1 with the angular and energy domain covered by the left block. The measured angular domain for the elastic cross sections is between 5-30$^\circ_{c.m}$. The (p,p') kinematics were obtained by the correlations of the proton energies with the scattering angle deduced from the proton trajectories measured in MUST2 and from the incident trajectories reconstructed using 3 PPACs located upstream of the target.

2.2. *Particle identification*

The incident particles were identified by Bρ-ETOF-ΔE technique in the (F3-F7) stages, and the ejectiles in the ZD [15] spectrometer. The rate of ^{24}O represented 4% of the total beams, including the light nuclei with A/Z=3 (t, ^{6}He, ^{9}Li). Amongst the heavy ion beams (Z > 3), it represented 22% of the total; contaminants were ^{20,21,22}N ($\simeq$30%), ^{22,23}O ($\simeq$20%), ^{25}F ($\simeq$12%) and ^{26}F ($\simeq$ 5%).

3. The analysis of the *Ex* spectra

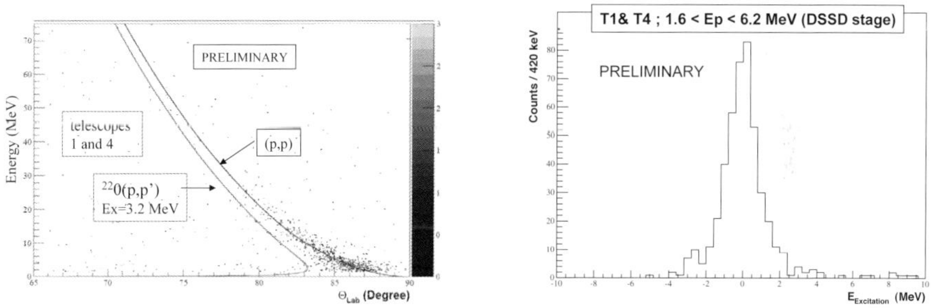

Fig. 2. The kinematics and Ex spectra for ^{22}O, showing the known 2_1^+ at 3.2 MeV.

To check the kinematical reconstruction of the reactions, the ^{22}O(p,p') at 262.5A.MeV were measured as a reference. The kinematics and the preliminary Ex spectra are shown in Fig. 2, the energy straggling effects in the target are taken into account.

The next steps are the extraction of the excitation spectra for ^{22}O and for its contaminants. The angular acceptance and detector efficiencies will be evaluated and compared with the estimates done using a Monte-Carlo simulation program. The elastic (p,p) angular distributions will be extracted and compared to model calculations. We will check the consistency of the data obtained for ^{22}O with the previous measurements done at GANIL at 46 MeV/n [16]. The analysis is also in progress for ^{24}O(p,p'). The data was collected with different settings of the magnetic rigidity tuned to center either the ^{23}O fragment or the ^{22}O at the focal plane of the spectrometer. For the event selection and the reconstruction of the ^{24}O Ex spectra we will take into account the correlations with the outgoing nucleus produced by the decay of the unbound states of ^{24}O, either ^{23}O (Ex above S_n) or the ^{22}O isotope (above S_{2n}=7.5 MeV).

Perspectives

The results will be compared with the recent microscopic calculations, including the interplay between cluster, single-particle and continuum-coupling effects [7,17,18]. From the characteristics of the excited states, we expect to deduce the properties of the N=16 shell gap in ^{24}O.

Acknowledgments

Special thanks are due to the accelerator team of the RIBF machine for delivering the ^{48}Ca at maximum intensities. RIKEN Nishina Center is gratefully acknowledged for supporting the stay of the French team during the campaign.

References

1. T. Otsuka *et al.*, *Phys. Rev. Lett.* **105**, 032501 (2010).
2. E. Pollacco *et al.*, *The MUST2 collaboration, Eur. Phys. J. A* **25**, 287 (2005).
3. X. Mougeot, V. Lapoux *et al.*, *submitted to Phys.Rev. Lett., June 2011.*
4. N. Keeley *et al.*, *Phys. Lett. B* **646**, 222 (2007).
5. S. C. Pieper, R. B. Wiringa, and J. Carlson, *Phys. Rev. C* **70**, 054325 (2004).
6. N. Michel, *et al.*, *Phys. Rev. C* **67**, 054311 (2003).
7. A. Volya and V. Zelevinsky, *Phys. Rev. Lett.* **94**, 052501 (2005).
8. G. Hagen, M. Hjorth-Jensen, J.S. Vaagen, *Phys. Rev. C* **71**, 044314 (2005).
9. *Evaluation of mass excess 2011*, Priv. Co. (Apr. 2011) G. Audi and W. Meng.
10. A. Ozawa *et al.*, *Phys. Rev. Lett.* **84**, 24 (2000).
11. T. Otsuka *et al.*, *Phys. Rev. Lett.* **91**, 179202 (2003).
12. S. Stanoiu *et al.*, *Phys. Rev. C* **69**, 034312 (2004).
13. C.R. Hoffman *et al.*, *Phys.Lett.* **672**, 17 (2009).
14. T. Kubo *et al.*, *NIM. Phys. Res. B* **204**, 97 (2003).
15. T. Ohnishi *et al.*, *J. Phys. Soc. Jpn.* **77**, 083201 (2008).
16. E. Becheva *et al.,MUST collaboration*, *Phys. Rev. Lett.* **96**, 012501 (2006).
17. K. Tsukiyama, M. Hjorth-Jensen, G. Hagen, *Phys. Rev. C* **80**, 051301 (2009).
18. K. Tsukiyama, T. Otsuka and R. Fujimoto, *arXiv:1001.0729 (2010).*

HIGH-RESOLUTION ION-OPTICAL ANALYSIS OF RI-BEAMS WITH THE SHARAQ SPECTROMETER

T. UESAKA*, S. MICHIMASA, H. TOKIEDA, S. SHIMOURA, S. OTA, Y. SASAMOTO, H. MIYA, S. KAWASE, Y. KIKUCHI, K. KISAMORI, M. TAKAKI, H. MATSUBARA, and M. DOZONO

Center for Nuclear Study, University of Tokyo
Bunkyo, Tokyo 113-0033, Japan
**Email: uesaka@cns.s.u-tokyo.ac.jp*
http://www.cns.s.u-tokyo.ac.jp/sharaq/

S. NOJI, K. MIKI, and K. YAKO

Department of Physics, University of Tokyo
Bunkyo, Tokyo 113-0033, Japan

H. SAKAI, H. TAKEDA, D. KAMEDA, T. OHNISHI, Y. YANAGISAWA, T. KUBO, and H. BABA

RIKEN Nishina Center,
Wako, Saitama 351-0198, Japan

T. KAWABATA

Department of Physics, Kyoto University
Kyoto 606-8502, Japan

P. Roussel-Chomaz

GANIL, DSM/CEA, IN2P3/CNRS, BP 5027, 14076 Caen Cedex 5, France

M. SASANO

National Superconducting Cyclotron Laboratory, Michigan State University
1 Cyclotron, East Lansing, Michigan 48824-1321, USA

G. P. A. Berg

Department of Physics and Joint Institute for Nuclear Astrophysics, University of Notre Dame,

225 Nieuwland Science Hall, Notre Dame, IN 46556-5670, USA

The high-resolution SHARAQ spectrometer has been constructed at the RI Beam Factory to facilitate missing mass spectroscopy with RI-beam induced reactions. Cathode read-out drift chambers developed under GANIL-CNS collaboration were installed at the SHARAQ focal plane and were successfully applied to tracking of $Z = 1$–7 ions. We studied ion-optical properties of the SHARAQ spectrometer and the beam-line and accomplished simultaneous fulfillment of lateral and angular dispersion-matching conditions for a secondary ^{10}C beam. A series of experiments to investigate IVMRs were performed with the SHARAQ spectrometer in 2009–2010.

Keywords: Magnetic spectrometer; Radioactive isotope beam; Charge exchange reactions.

1. High-resolution ion-optical analysis of RI-beams

Study of spin-isospin responses in nuclei forms one of cores in nuclear physics.[1] A variety of collective states, for example isovector giant dipole resonances, isobaric analogue states, Gamow-Teller resonances, have been extensively studies by use of electromagnetic and hadronic reactions. Despite decades of efforts, however, several states of fundamental interests are left to be discovered. They include double Gamow-Teller resonances (DGTR), β^+-type isovector spin monopole resonances, 0^- resonances, and so on. Their absences are mostly due to lack of proper experimental ways to approach the states.

RI-beam induced charge exchange reactions (RICE) have unique properties which are missing in stable-beam induced reactions and can be used to reach the yet-to-be-discovered states. High-resolution ion-optical analysis of RI-beams is a critical technique in use of RICE for the spectroscopic purposes, . We have constructed the high-resolution SHARAQ spectrometer[2] at the RI Beam Factory to pursue the capabilities of RICE as new probes to nuclei.

2. SHARAQ spectrometer and Cathode Read-out Drift Chamber

The SHARAQ spectrometer was constructed to facilitate missing mass spectroscopy with RICE at 100–300 MeV/nucleon. Here, high-momentum and high-angular resolutions are crucial. The SHARAQ spectrometer is of a *QQDQD* configuration and is designed to achieve a momentum resolution of $\delta p/p$=1/14700 for particles with a magnetic rigidity of 1.8–6.8 Tm. The momentum resolution corresponds to an excitation energy resolution of $\sim$ 0.5 MeV in a reaction with a projectile with a mass number of 10 and an incident energy of 200 MeV/nucleon. It is designed also to achieve an angular resolution of $\delta\theta \sim 1$ mrad, which corresponds to a momentum transfer resolution of 10 MeV/c.

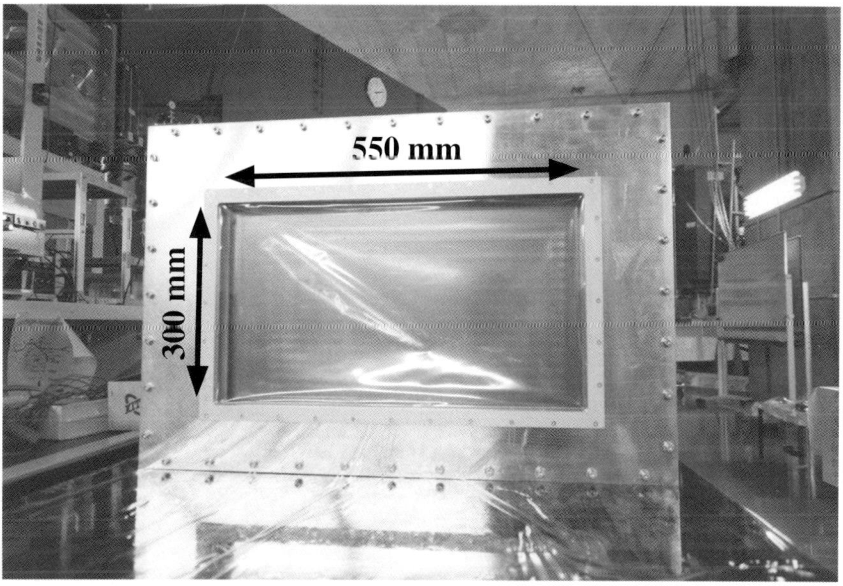

Fig. 1. Photograph of CRDC.

Momentum and scattering angles of the ejectile are reconstructed from trajectory at the SHARAQ focal plane. To achieve the required momentum and angular resolutions, the positions and angles of the trajectory should be determined with a uncertainty of 0.3 mm and 1 mrad in FWHM.

Cathode read-out drift chambers at the SHARAQ focal plane have been constructed under collaboration of CNS and GANIL (see Fig. 1). The CRDC is designed to operate with a low pressure isobuthane gas of a few kPa[3] to minimize multiple scattering which may ruin the angular resolution. Two CRDCs with an effective area of 550×300 mm^2 are placed in an interval of 500 mm to cover an acceptance of the spectrometer.

In the commissioning runs carried out in 2009, the CRDCs were tested for light ions of $Z = 1$–7. They were irradiated with a ^{14}N beam at 250A MeV and ^{12}B, ^{9}Li, ^{6}He, and ^{3}H beams at 200A MeV. Detection efficiencies were found to be almost 100% at anode high voltages of 650,

800, 850, 950 V for ^{14}N, ^{9}Li, ^{6}He, and ^{3}H, respectively. Horizontal position resolution is as good as 0.42 mm (FWHM) for each plane, resulting in a resolution of the focal plane position of $\Delta x = 0.3$ mm, which meets the design requirements.

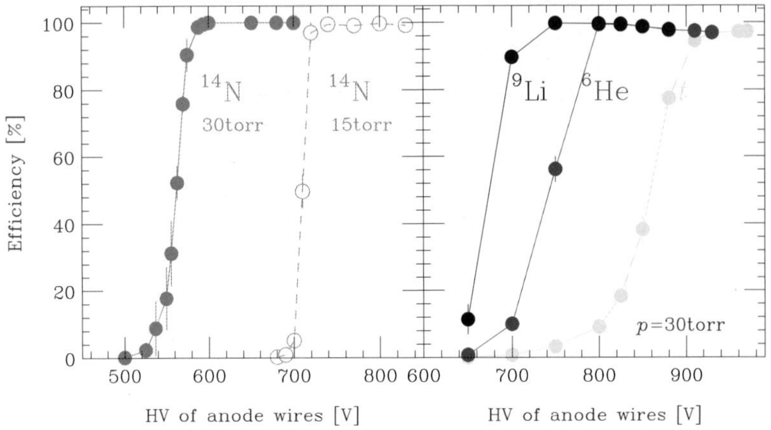

Fig. 2. Detection efficiencies for $Z = 1$–7.

3. Dispersion matching of secondary beam

The beam-line is designed to be dispersion-matched to the spectrometer[4] for accomplishing the required resolutions for a secondary RI beams with large momentum spread. In October 2010, we have tested the SHARAQ+beamline system for secondary ^{10}C and ^{12}N beams and succeeded in satisfying lateral and angular dispersion matching conditions simultaneously. Figure 3 shows correlations between horizontal position and angle of particle trajectories at the focal plane to the horizontal position at a dispersive focal plane (FH6) in the beam-line after beam tuning. The focal plane FH7 has a momentum dispersion of 7.3 m and the position measured there is a good measure of beam momentum. Upright correlations shown in Fig. 3 means that the position and the angle at the SHARAQ focal plane are independent from the beam momentum, which in turn means a simultaneous fulfillment of lateral and angular dispersion matching conditions.

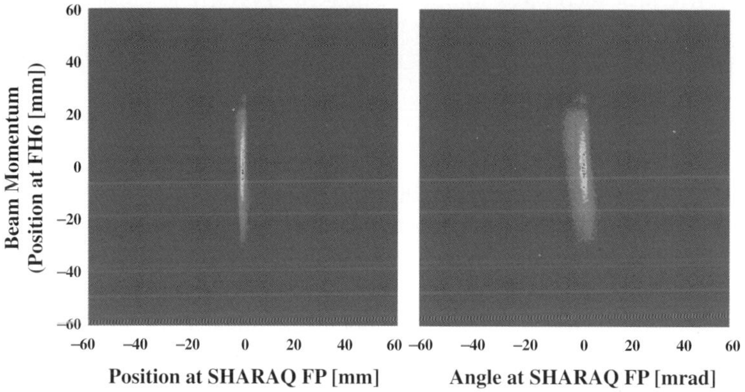

Fig. 3. Correlations of particle trajectories at the SHARAQ focal plane and the dispersive focal plane (FH6) in the beam-line.

4. Physics cases

A series of experiments were conducted in 2009–2010 to investigate isovector monopole resonances (IVMR) in ^{90}Zr and ^{208}Pb. IVMRs are isovector analogues of isoscalar monopole resonances which have strong relevances to incompressibility of nuclei,[5] and thus properties of IVMR are expected to give impacts on nuclear matter theories.[6]

The first experiments with the SHARAQ spectrometer was performed to search for β^+-type isovector spin monopole resonances (IVSMR). The $(t, {}^3\mathrm{He})$ reaction at 300 MeV/nucleon was used to extract β^+-strengths selectively. An intense triton beam of 10^7 sec^{-1} was produced by the projectile fragmentation of a primary 320 MeV/nucleon ^{4}He beam and the scattered ^{3}He ions were momentum-analyzed by the SHARAQ spectrometer. Clear signatures of the IVSMR were observed, for the first time, at excitation energies of 10–30 MeV in ^{90}Zr and ^{208}Pb, respectively.[7] The values of excitation energies are in good agreement with HF+Tamm-Dancoff calculations.[8]

In 2010, two experiments were carried out to reveal different aspects of IVMR: one is the $^{90}\mathrm{Zr}({}^{10}\mathrm{C}, {}^{10}\mathrm{B}(0^+))$ experiment to search for isovector non-spin-flip monopole resonances. The reaction selectively populates isovector (ΔT=1) non-spin-flip (ΔS=0) modes in nuclei,[9] while any stable-beam induced charge exchange reaction never has the selectivity. In the other experiment, we explored a new method based on the exothermic charge exchange $({}^{12}\mathrm{N}, {}^{12}\mathrm{C})$ reaction.[?] The reaction has a large positive Q-value of about +17 MeV, which enables us to populate the highly-excited

β^--IVSMR in a recoil-less manner. Since β^--IVSMR is most strongly excited at around q=0, prominent strength is expected to be observed by the exothermic reaction.

5. Summary

We have constructed the high-resolution SHARAQ spectrometer at the RI Beam Factory to pursue the capabilities of RICE as new probes to nuclei. CRDCs fabricated under GANIL-CNS collaboration were successfully applied to tracking of $Z = 1$–7 ions. By using particle trajectories determined by CRDCs and beam-line detectors, we accomplished simultaneous fulfillment of lateral and angular dispersion-matching conditions for a secondary ^{10}C beam. A series of experiments to investigate IVMRs were performed with the SHARAQ spectrometer in 2009–2010.

Acknowledgements

The present work was partly supported by the Grant-in-Aid for Scientific Research (Nos. 17002003 and 19204024) by the Ministry of Education, Culture, Sports, Science and Technology. The authors thank JSPS Core-to-Core program.

References

1. M. N. Harakeh and A. V. D. Woude, *Giant Resonances: Fundamental High-Frequency Modes of Nuclear Excitation* (Oxford University Press, 2001).
2. T. Uesaka et al., *Nuclear Instruments and Methods in Scientific Research B* **266**, 4218 (2008).
3. S. Michimasa et al., P. Roussel-Chomaz, J.-F. Libin, P. Gangnant and C. Spitaels, *RIKEN Accelerator Progress Report* **42**, 177 (2009).
4. T. Kawabata, et al., *Nuclear Instruments and Methods in Scientific Research B* **266**, 4201 (2008).
5. J. Blaizot, *Physics Reports* **64**, 171 (1980).
6. G. Colo et al., *Physical Review C* **67**, 044306 (2003).
7. K. Miki et al., in *Proceedings of International Nuclear Physics Conference 2010, in press.*
8. I. Hamamoto and H. Sagawa, *Physical Review C* **62**, 024319 (2000).
9. Y. Sasamoto et al., The super-allowed fermi type charge exchange reaction for studies of isovector non-spin-flip monopole resonance, in *Proceedings of French-Japanese Symposium on Nuclear Structure Problems*, 2011.
10. S. Noji et al., Proposal for Nuclear Physics Experiment at RI Beam Factory, "Development of the exothermic charge-exchange reaction as a new spin-isospin probe and its application to the observation of the isovector spin monopole resonance in ^{90}Nb", 2009.

Nuclear moments of μ-second isomeric fragments at BigRIPS

R. CHEVRIER, J. M. DAUGAS, L. GAUDEFROY

CEA, DAM, DIF, 91297 Arpajon, France

M. HASS

Departement of Particle Physics, Weizmann Institute of Science, Rehovot 76100, Israel

H. HAAS

EP Division, CERN, 1211 Geneva, Switzerland

H. UENO, N. AOI, N. FUKUDA, Y. ICHIKAWA, N. INABE, M. ISHIHARA, D. KAMEDA, T. KUBO, T. OHNISHI, H. TAKEDA, H. WATANABE, A. YOSHIMI

RIKEN Nishina Center, RIKEN, 2-1 Hirosawa, Wako, Saitama 351-0198, Japan

K. ASAHI, T. FURUKAWA, H. HAYASHI, H. IIJIMA, T. INOUE, Y. ISHII, T. NANAO, K. SUZUKI, M. TSUCHIYA

Department of Physics, Tokyo Institute of Technology, 2-12-1 Oh-okayama, Meguro-ku, Tokyo 152-8551, Japan

D. L. BALABANSKI

INRNE, Bulgarian Academy of Sciences, 1784 Sofia, Bulgaria

G. GEORGIEV

CSNSM, Université Paris-Sud 11, CNRS, IN2P3, 91405 Orsay-Campus, France

S. COOTENIER

Center of Molecular Modeling, Ghent University, Technologiepark 903, 9052 Zwijnaarde, Belgium

G. NEYENS, M. RAJABALI

Instituut voor Kern- en Stralingsfysica, K.U. Leuven, 3001 Leuven, Belgium

The measurement of nuclear moments of a μ-second isomeric state have been successfully performed using the RIKEN Radioactive Ion Beam Facility. The Time Dependent Perturbed Angular Distribution technique was applied on the ^{43}S isomeric state produced by the fragmentation of a ^{48}Ca beam and selected with the BigRIPS spectrometer. Following the measurement of the known g-factor of the isomer the quadrupole moment of this state was measured. Its spectroscopic quadrupole moment $|Q_S| = 23(3)$ efm^2 agreed with both shell model and beyond mean field calculations.

Keywords: Electromagnetic moments; Projectile fragmentation; Nuclear structure.

1. Introduction

One of the major preoccupation of the experimental and theoretical nuclear physics nowadays is the identification of major proton and neutron shell closures far from the valley of β-stability. The understanding of the behavior of the structure of orbitals filled by the valence nucleons for more and more exotic nuclei, up to the drip lines, will allow a better description of the interaction force applied between nucleons. The evolution of the orbital energies with the isospin, the appearance of new magic number and the weakening of the well known spherical closed shells were studied for light neutron-rich nuclei.

Nuclear moments are very sensitive probes to the detailed composition of the nuclear wave function. The magnetic moment of a nuclear state provides unique information on its single particle structure and is a test ground for nuclear models. Due to its dependence on the spin and orbital angular momentum of the involved valence nucleons, the magnetic moment is a rigorous probe for the spin and parity assignment of the nuclear states. The electric quadrupole moment is a measure of the shape and the ensuing deformation of the corresponding nuclear level. As such, it provides unique information on the collective properties of the investigated states.

Recent intense experimental and theoretical interest was devoted to the study of neutron-rich nuclei around $N \approx 28$.[1–7] The spherical $N = 28$ shell gap erodes when moving away from the stable isotopes giving rise to deformation shape.[8–12] Particularly, shape coexistence is predicted along the $N = 28$ neutron rich isotopes[8] and observed for the $Z = 16$ ^{44}S.[3,11] Evidence of this coexistence was also provided by the measurement of the magnetic moment of the isomeric state $t_{1/2} = 415(5)$ ns in the ^{43}S at an excitation energy of 320 keV.[13] The obtained g-factor, $g = -0.317(4)$, clearly assigned this state from the normal $\nu f_{7/2}$ orbital with a $7/2^-$ spin-parity and the intruder nature of the ground state providing direct and unambiguous evidence of the collapse of the N=28 shell closure in neutron-rich nuclei.

We report in this proceeding the validation of the method for the measurement of magnetic moment of μs isomeric state at RIKEN with the Radioactive Ion Beam Facility (RIBF) and the measurement of the quadrupole moment of the ^{43}S isomer.

2. Experiment

Nuclear moment measurements of μs isomers are performed using the Time Differential Perturbed Angular Distribution (TDPAD)[14] method in combination with ion-γ correlations. For the magnetic moment determination the isomers are stopped in a perturbation free environment by choosing an appropriate stopper material that has cubic lattice structure. The implantation foil is placed between the poles of an electromagnet, delivering an external magnetic field B, and causing a Larmor precession of the initially aligned isomeric nuclear spins with a frequency

$$\omega_L = -\frac{g\mu_N B}{\hbar}, \tag{1}$$

where g is the nuclear gyromagnetic factor; μ_N the nuclear magneton and $\hbar$ Plank's constant. For the quadrupole moment measurement the isomers are implanted in a crystal which have an internal electric field gradient (EFG). The resulting quadrupole interaction frequency could be expressed as a superposition of several harmonics with basic frequency:

$$\omega_0 = \frac{neQ_S V_{ZZ}}{4I(2I-1)\hbar}, \tag{2}$$

where I the spin, $n = 3(6)$ for integer (half-integer) spin, Q_S the spectroscopic quadrupole moment, and V_{ZZ} the EFG.

For both measurements the monitoring of γ radiations is performed using 4 high-purity Ge detectors. For each Ge detector, energy and time signals are stored in an event by event mode. γ-ray time spectra are started by the ion implantation time, and stopped by the detection of a delayed γ-ray. To extract the precession pattern out of the individual time spectra, detectors at 90° with respect to each other are combined to generate the $R(t)$-function:

$$R(t) = \frac{I_1(\theta,t) - \epsilon I_2(\theta+\frac{\pi}{2},t)}{I_1(\theta,t) + \epsilon I_2(\theta+\frac{\pi}{2},t)}, \tag{3}$$

where I_1 and I_2 are the summed intensities of detectors placed at 180°; ϵ the relative efficiency between detectors; θ the angle between the beam axis and the first detector.

Nuclei of interest were produced by the fragmentation of a ^{48}Ca beam at an energy of 345 AMeV with an mean intensity of 200 pnA impinging a 4 mm ^{9}Be target. Fragments were selected using the BigRIPS spectrometer[15] and a wedge degrador foil placed at the first dispersive plane of the spectrometer. The detection system was located at the F7 focal plane. The time of flight of the fragments through the spectrometer was about 350 ns. The experimental setups for the magnetic moment and for the quadrupole moment measurements are presented in Fig. 1 left hand panel and right hand panel, respectively. At the target position the initial spin-alignment axis is located along the spectrometer axis. At the detection setup position it is possible to determined, knowing the g-factor, the orientation axis which is at an angle $\alpha = \theta_C(1 - \frac{gA}{2Q})$, where θ_C is the rotation angle in the spectrometer; A the mass of the implanted ion and Q its charge. For the ^{43m}S, $\alpha = 86°$. According to this angle and knowing that, in order to have the better asymmetry of the $R(t)$ function, the Ge detectors has to be located along and perpendicular to the orientation axis we put the detectors as shown in the right hand panel of Fig. 1.

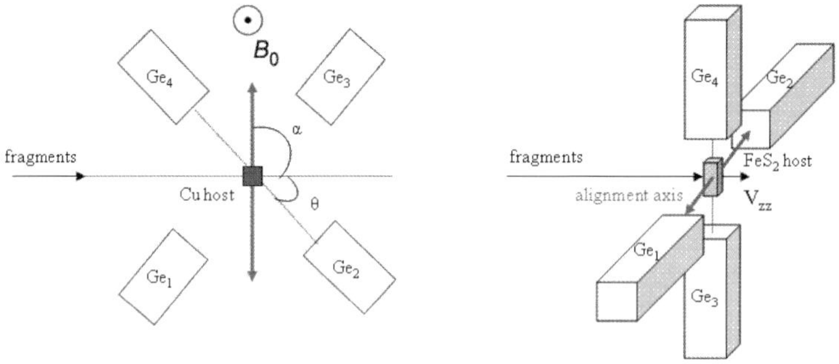

Fig. 1. Schematic drawing of the experimental setup for the present g-factor (left panel) and quadrupole moment (right panel) measurements. Angles are defined in the text.

3. Results

For the proof of principle of using the TDPAD technique with fragments produced with the BigRIPS spectrometer and to perform a selection where the amount of spin-alignment is the highest, we have first measured the g-factor of the ^{43m}S at two selections in the momentum distribution, as shown in Fig.2 left panel. The measurement has been performed using a Cu host and an applied magnetic field $B = 0.670(1)$ T. An amount of spin-alignment of $+8.1(9)\%$ and $-2.8(6)\%$ is observed in the center (Fig. 2 center panel) and in the outer wing (Fig. 2 right panel) of the momentum distribution, respectively. The measured $g = -0.312(12)$ is in agreement with the previous measurement.[13]

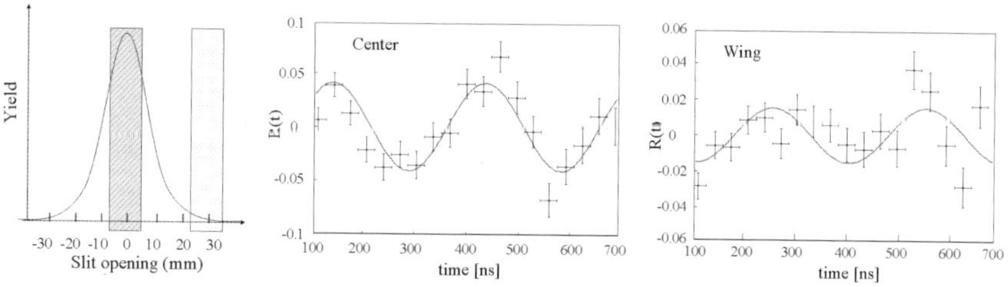

Fig. 2. g-factor measurement at different selections in the momentum distribution (left panel). The center panel represents the selection in the center and the right hand panel the selection in the outer wing.

A pyrite (FeS_2) single crystal has been used to measure the quadrupole moment. The calculated EFG at S position is predicted to be $+14.14 \times 10^{21}$ V/cm^2 within 10% reliability. The $R(t)$ function of the quadrupole moment measured in the center of the momentum distribution is presented in Fig. 3. The observed oscillation pattern leads to a spectroscopic quadrupole moment of $|Q_S| = 23(3)$ efm^2. Shell model calculations per-

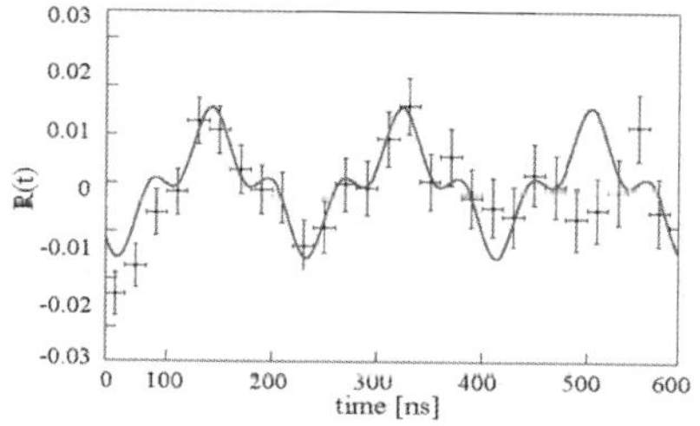

Fig. 3. $R(t)$ function corresponding to the quadrupole moment measurement.

formed using the SDPF-U interaction[1] give a spectroscopic quadrupole moment of $Q_S = +25$ efm^2 in good agreement with our results. A value of $Q_S = -13$ efm^2 is predicted for the ground state. Mean field and beyond mean field calculations using the Generator Coordinate Method (GCM) implemented with the Gaussian Overlap Approximation (GOA)[16] performed using the Gogny D1S force[17] predict oblate $\beta = -0.21$ $K^\pi = 1/2^-$ state and prolate $\beta = +0.15$ $K^\pi = 7/2^-$ state with an identical spectroscopic quadrupole moment $Q_S = +14$ efm^2, slightly underestimated the deformation. These calculations predict a prolate deformation $\beta = +0.35$ and a spectroscopic quadrupole moment $Q_S = -15$ efm^2 for the $3/2^-$ ground state originate from the intruder $\nu p_{3/2}$ orbital.

The present measurement leads to a moderated prolate deformed isomeric state $\beta = +0.24(3)$ or a more deformed oblate state $\beta = -0.33(4)$. The measurement of levels above the isomer might give useful information to determined the shape of this state.

Nuclear moments of a μs isomeric state have been measured using the RIBF at RIKEN. These give the opportunity to investigate very exotic isomers and to study the evolution of the shell structure of nuclei approaching the drip lines.

Acknowledgments

We thank the technical staff of the accelerator facility and the BigRIPS spectrometer at RIKEN. This work was supported by the Japanese-French Hubert Curien partnership (PHC). It is implemented by the Japan Society for the Promotion Science (JSPS) and by the Ministére de l'Enseignement Supérieur et de la Recherche (MESR) and by the Ministére des Affaires Etrangéres et Européennes (MAEE).

References

1. F. Sarazin *et al.*, *Phys. Rev. Lett.* **84**, 3082 (2000).
2. D. Sohler *et al.*, *Phys. Rev. C* **66**, 054302 (2002).
3. S. Grévy *et al.*, *Eur. Phys. J. A* **25**, Supplement 1, 111 (2005).
4. F. Nowacki and A. Poves, *Phys. Rev. C* **79**, 014310 (2009).
5. R. Ringle *et al.*, *Phys. Rev. C* **80**, 064321 (2009).
6. L. Gaudefroy, *Phys. Rev. C.* **81**, 064329 (2010).
7. K. Kaneko, Y. Sun, T. Mizusaki, M. Hasegawa, *Phys. Rev. C* **83**, 014320 (2011).
8. S. Péru, M. Girod and J.F. Berger, *Eur. Phys. J. A* **9**, 35 (2000).
9. B. Bastin *et al.*, *Phys. Rev. Lett.* **99**, 022503 (2007).
10. M. Zielinska *et al.*, *Phys. Rev. C* **80**, 014317 (2009).
11. C. Force *et al.*, *Phys. Rev. Lett.* **105**, 102501 (2010).
12. L. Caceres *et al.*, *Acta Phys. Pol. B* **42**, 533 (2011).
13. L. Gaudefroy *et al.*, *Phys. Rev. Lett.* **102**, 092501 (2009).

14. G. Goldring and M. Hass, *Treaties in Heavy Ion Sciences*, D. E. Bromley ed. Plenum Press, Vol. 3, p.539.
15. T. Kubo, *Nucl. Instr. Meth. B* **204**, 97 (2003).
16. J. Libert, M. Girod and J. P. Delaroche, *Phys. Rev. C* **60**, 054301 (1999).
17. J. Dechargé and D. Gogny, *Phys. Rev. C* **21**, 1568 (1980).

Production of spin-aligned RI beams via the two-step fragmentation reaction

H. Ueno[1], Y. Ichikawa[1], Y. Ishii[2], T. Furukawa[2], A. Yoshimi[1], D. Kameda[1], H. Watanabe[1], N. Aoi[1], K. Asahi[2], D.L. Balabanski[3], R. Chevrier[4], J. M. Daugas[4], N. Fukuda[1], G. Georgiev[5], H. Hayashi[2], H. Iijima[2], N. Inabe[1], T. Inoue[2], M. Ishihara[1], T. Kubo[1], T. Nanao[2], T. Ohnishi[1], K. Suzuki[2], M. Tsuchiya[2], H. Takeda[1], M.M. Rajabali[6]

[1]*RIKEN Nishina Center, 2-1 Hirosawa, Wako, Saitama 351-0198, Japan*
[2]*Department of Physics, Tokyo Institute of Technology, 2-12-1 Oh-okayama, Meguro, Tokyo 152-8551, Japan*
[3]*Institute for Nuclear Research and Nuclear Energy, Bulgarian Academy of Sciences, BG-1784, Sofia, Bulgaria*
[4]*CEA, DAM, DIF, F-91297 Arpajon, France*
[5]*CSNSM, IN2P3-CNRS, Université Paris-sud, F-91404 Orsay, France*
[6]*Instituut voor Kern- en Stralingsfysica, K.U. Leuven, Celestijnenlaan 200D, B-3001 Leuven, Belgium*

Based on the technique of fragmentation-induced spin polarization combined with the β-NMR method, we have conducted a series of ground-state nuclear-moment measurements in a region of light neutron-rich nuclei at RIKEN. The research has recently been extended also at GANIL. The ground-state studies will be further enhanced by the upgrade of RIPS proposed in the RIBF Phase-II program. Also, utilizing the technique of fragmentation-induced spin-orientation, nuclear-moment measurements of isomeric state were conducted. Remarkably, we have recently developed a method to produce highly spin-aligned RI beams in a new promising scheme. The method was first examined using BigRIPS, where a highly spin-aligned isomeric ^{32m}Al beam was successfully produced through the two-step projectile fragmentation reaction.

Keywords: Electromagnetic moments; Polarized beams; Projectile fragmentation.

1. Introduction

The RIKEN RI Beam Factory (RIBF) project has started its operation since 2007,[1] where the superconducting in-flight radioactive-isotope beam (RIB) separator BigRIPS[2] and the following ZeroDegree, SHARAQ, and SAMURAI spectrometers[3] have been installed together with the three new ring cyclotrons fRC, IRC and SRC. The RIBF facility is designed to provide various kinds of RIBs with the world's highest intensities based on the in-flight method. Making use of such spectrometers, more than 20 experimental programs have already been conducted with intense beams delivered by the RIBF cyclotron cascade acceleration. In addition to these reaction-based spectroscopic studies, the EURICA (Euroball RIKEN Cluster Array) project,[4] a stopped-RIB campaign at RIBF with the EUROBALL Ge detectors for γ-decay spectroscopy, will unfold over the period of year and half starting from April, 2012.

Besides such the spectroscopic approaches, studies with spin-oriented RIBs are also ongoing at RIBF. For the measurement of ground-state nuclear moments, fragment-induced spin-polarized radioactive-isotope beams (RIBs)[5,6] offers us the opportunity of studying on the structures of nuclei far from the stability line, through the measurement of electromagnetic nuclear moments. Based on this technique, a series of ground-state nuclear-moment measurements has been carried out by means of the β-ray detected nuclear magnetic resonance (β-NMR) method with the former intermediate-energy fragment separator RIPS.[7] In addition, the production of a spin-aligned RIB was successfully produced with BigRIPS.[2] It should be noted that in the study of spin-aligned RIBs, we have developed a method to produce highly spin-aligned RI beams in a new promising scheme.

2. Spin-oriented RI beams and nuclear-moment measurements

It has been revealed in our earlier work that spin oriented RIBs can be produced in the projectile-fragmentation (PF) reaction at intermediate energies.[5,6] Utilizing thus obtained fragmentation-induced spin-polarized RIBs, a series of nuclear-moment measurements have been carried out in the neutron-rich p- and sd-shell nuclear region in RIKEN,[8] where 17 new nuclear moments have been determined so far. The obtained experimental nuclear moments were found to be useful to discuss, for instance, the spin-parity assignment of ^{17}C,[9] effects of excess neutrons,[10] the reduction of effective charges,[11] and the structure of nuclei near the "island of inversion".[12–14] With respect to the study on neutron-rich aluminum isotopes, their magnetic-moment measurements were conducted in GANIL[15–17] independently of the RIKEN group so far. Recently, we have extended the Q-moment measurement of neutron-rich aluminum isotopes at GANIL in collaboration with the GANIL group.[18,19]

For excited states of RIBs, spin-aligned phenomena found in the PF reaction[20,21] have been utilized. This method is suitable for the isomeric-state nuclear-moment measurement of RIBs in view of the production yield of RIBs, the population of isomeric states, and production of spin-aligned RIBs. Taking these advantages, nuclear-moment measurements of unstable nuclei have already been carried out at GANIL[22] in combination with the time differential perturbed angular distribution (TDPAD) method.

3. New scheme to produce spin-aligned RI beams

Although there are are many advantages in this method, it should be taken into account that spin orientation produced in the PF reaction is known to be very small or disappeared for fragments implying removals of many nucleons from the projectile. For instance, spin polarization of a ^{31}Al produced in the PF of a ^{40}Ar beam[13] was found to be one order smaller than that produced in the fragmentation of ^{36}S.[18] This observation indicates that PAD measurements are limited only near primary beams.

In order to overcome this difficulty, we planned to utilize the two-step PF reaction method[23] for the production of spin-aligned RIBs, where a secondary beam produced in the primary PF is used to yield an RIB of interest in the secondary PF reaction as a *tertiary* RI beam. In this scheme, a secondary beam is selected so that a *tertiary* RI beam can be produced in one-nucleon removal fragmentation. Then, a *tertiary* RI beam can be *promisingly* spin-aligned as high as the PF-induced method can produce, since the mechanism of the spin alignment is purely described as given in Ref.[20] due to the reaction simplicity. Also, we note that large enhancement of the total production yield was suggested in the in-beam γ-ray spectroscopy.[23]

4. Production of a spin-aligned ^{32m}Al beam

The first experiment to produce a spin-aligned RI beam via the two-step fragmentation was conducted with the BigRIPS in-flight separator[2] at the RIKEN RIBF facility.[1] In the experiment, ^{33}Al was first produced as a secondary beam in the fragmentation of a ^{48}Ca beam on a ^{9}Be target at $E/A = 345$ MeV so as to provide a maximum production yield. The secondary ^{33}Al beam was introduced to an aluminum wedge-shaped secondary target, placed at the momentum dispersive focal plane, in order to produce ^{32}Al as a *tertiary* beam through one-nucleon removal PF reaction. The isomeric state ^{32m}Al [6] was populated and the spin-aligned effectively at the same time in a *one-shot* reaction. The ^{32m}Al *tertiary* beam was, then, introduced and implanted into a Cu crystal mounted in a TDPAD apparatus located at an achromatic focal plane of BigRIPS. The magnitude of the ^{32m}Al spin alignment was determined by means of the TDPAD method, where the spin-alignment was observed through the change in the anisotropy of the de-excitation γ-ray emitted from the implanted ^{32m}Al nuclei, measured with four Ge detector placed surrounding the stopper crystal, under an external magnetic field.

In a preliminary analysis, large spin alignment $\sim$ 10 % was obtained for ^{32m}Al in spite of the large nucleon removal $\Delta A = 16$ from the ^{48}Ca primary beam, which corresponds to 33 % mass removal. For comparison, spin alignment of ^{32m}Al was also measured for a conventional single-step scheme, where ^{32m}Al was directly produced from ^{48}Ca, where no significant spin alignment was observed with the almost same statistics, suggesting small spin alignment $\leq$ 2 %. Detailed description of the experiment and data analysis will be given in Ref.[24]

5. Beam-transport line from IRC to RIPS

In the RIBF new facility, we plan to conduct nuclear-moment measurements not only for isomeric states, *e.g.*, through the new two-step fragmentation scheme, but also for ground states through the following new program.

The RIBF project will be fully capitalized in the phase-II program, in which the construction of some several experimental key devices[3] have been already completed. An upgrade of RIPS[7] has been also proposed as one of the phase-II program. In the cyclotron-cascade acceleration scheme, beams are accelerated up to the energy of $E = 115$ A MeV with IRC. In this upgrade, the former fragment separator RIPS is equipped with this new beam transport line[25] that delivers beams of 115 A MeV heavy ions from the IRC cyclotron. RI beams produced by the primary beams at such an intermediate energy are high enough to produce RIBs via PF reactions and suitably low in energy to be stopped in a sample material of limited thicknesses. Remarkably, RIBs produced at $E/A = 115$ MeV can be spin-oriented by the mechanism mentioned in the above. This upgrade will provide an opportunities for further investigations of ground-state nuclear moments in the heavier mass region.

6. Summary

Based on the technique of fragmentation-induced spin polarization combined with the β-NMR method, we have conducted a series of ground-state nuclear-moment measurements in a region of light neutron-rich nuclei at RIKEN. The research has recently been extended also at GANIL, where static electric quadrupole moments of neutron-rich aluminum isotopes $^{31,\,33}$Al were measured. The ground-state studies will be further enhanced by the upgrade of RIPS proposed in the RIBF Phase-II program. Also, utilizing the technique of fragmentation-induced spin-orientation, nuclear-moment measurements of isomeric state were conducted using BigRIPS. Remarkably, we have recently developed a method to produce highly spin-aligned RI beams in a new *promising* scheme. The method was first examined using BigRIPS, where a highly spin-aligned isomeric ^{32m}Al beam was successfully produced through the two-step projectile fragmentation reaction. We plan to apply this method to nuclei far from the stability line as the next step.

References

1. Y. Yano, *Proc. 17th Int. Conf. on Cyclotrons and their Applications*, Tokyo Japan, October 18-22, 2004, eds. A. Goto and Y. Yano, Particle Accelerator Society of Japan, 169 (2005).
2. T. Kubo, Nucl. Instr. Meth. B **204**, 97 (2003).
3. Technical information on experimental installations are found at `http://ribf.riken.jp/RIBF-TAC05`.
4. Project information are found at `http://ribf.riken.jp/EURICA`.
5. K. Asahi *et al.*, Phys. Lett. B **251**, 488 (1990).
6. H. Okuno *et al.*, Phys. Lett. B **335**, 29 (1994).
7. T. Kubo *et al.*, Nucl. Instr. Meth. B **70**, 309 (1992).
8. K. Asahi *et al.*, Nucl. Phys. A **704**, 88c (2002) and references theirin.
9. H. Ogawa *et al.*, Eur. Phys. J. A **13**, 81 (2002)
10. H. Ueno *et al.*, Phys. Rev. C **53**, 2142 (1996)
11. H. Ogawa *et al.*, Phys. Rev. C **67** 064308 (2003)
12. H. Ueno *et al.*, Phys. Lett. B **615**, 186 (2005).
13. D. Nagae *et al.*, Phys. Rev. C **79** (2009) 027301.
14. D. Kameda *et al.*, Phys. Lett. B **647**, 93 (2007).
15. D. Borremans *et al.*, Phys. Lett. **B537** (2002) 45.
16. P. Himpe *et al*, Phys. Lett. **B643** (2006) 257.
17. P. Himpe *et al.*, Phys. Lett. **B658** (2008) 203.
18. M. De Rydt *et al.*, Phys. Lett. **B678** (2009) 344.
19. Shimada *et al.*, *Submitted to* Phys. Lett. **B** (2011).
20. K. Asahi *et al.*, Phys. Rev. C **43**, 456 (1991).
21. W.-D. Schmidt-Ott *et al.*, Z. Phys. A **350**, 215 (1994)
22. G. Georgiev *et al.*, J. Phys. G **28**, 2993 (2002); I. Matea *et al.*, Phys. Rev. Lett. **93** 142503 (2004).
23. K. Yoneda *et al.*, Phys. Lett. B **499**, 233 (2001)
24. Y. Ichikawa *et al.*, *under preparation*.

25. H. Ueno *et al.*, Eur. Phys. J. Special Topics **150**, 185 (2007).

RECENT STUDIES OF TRANSFER REACTIONS WITH MUST2 AT GANIL AND RIKEN

D.BEAUMEL
Institut de Physique Nucléaire, Université Paris-Sud, In2P3-CNRS
F-91406 Orsay cedex, France

Shell structure evolution in the regions around Z=8 and around N=68 has been inferred from transfer reaction studies using the MUST2 detector. A low-lying state in ^{12}O has been observed in the study of the two-neutron transfer reaction ^{14}O(p,t) evidencing mirror symmetry of the previously observed quenching of N=8 magic number on the neutron-rich side. In the higher mass region the ^{68}Ni(d,p) reaction has been investigated to selectively study the neutron shell orbitals that are crucial to interpret the strong shell effects observed around ^{68}Ni. Preliminary results show a strongly excited state around 2.5 MeV excitation energy in ^{69}Ni that could correspond to the 2d5/2 orbital.

1. Mirror symmetry of shell gap quenching in light nuclei

Strong shell structure changes have been observed in light neutron-rich nuclei. A often-quoted example is the case of the N=7 isotones. The N=8 shell energy gap represented by the E(s $_{1/2}$)-E(p $_{1/2}$) strongly decreases from ^{15}O to ^{11}B, and becomes negative in ^{11}Be, which ground-state's J^π is equal to 1/2+ instead of the normally expected 1/2- value. This so-called "parity inversion" is responsible of the halo phenomenon, being associated with a weak binding of the last neutron. Beyond the drip line, the parity inversion has been claimed in ^{10}Li, and indications have been brought for ^{9}He, still subject of investigations. The validity of the mirror symmetry of these effects at extreme conditions of isospin and binding energies remains an open question. On the proton-rich side, recent studies of ^{11}N have revealed that its ground state has a spin-parity of $1/2^+$ [1,2].
In the present work, mirror symmetry in the shell quenching phenomena between $^{12}O_4$ and its mirror partner $^{12}Be_8$ is investigated experimentally from the low-lying excitation properties. We have studied the structure of ^{12}O via missing-mass spectroscopy applied to the ^{14}O(p,t) reaction at 51 A.MeV. The systematics of the low-lying excited states in even-even nuclei provides a sensitive probe for the evolution of the shell structure. In ^{12}Be, the anomalously low excitation energies of the 2^+ [3], 1^- [4], and 0^+ [5] excited states, clearly

indicate the reduced shell gap at N=8 between the p and sd shells. For ^{12}O, however, experimental difficulties have hampered establishing the level scheme. The experiment was performed at the GANIL facility. The secondary ^{14}O beam at 51 A.MeV was produced by fragmentation of ^{16}O at 90 A.MeV on a 920-mg/cm^2-thick C target. A measurement was also performed with a degraded ^{16}O beam at 39~AMeV to obtain reference data. The energies and angles of the recoiling tritons were measured by an array of four MUST2 telescopes located 30~cm downstream of the target. The ejectiles were detected by SPEG near zero degrees and by a set of Si ΔE-E telescopes provided by RIKEN at larger angles. A detailed description of the setup is given elsewhere [6].

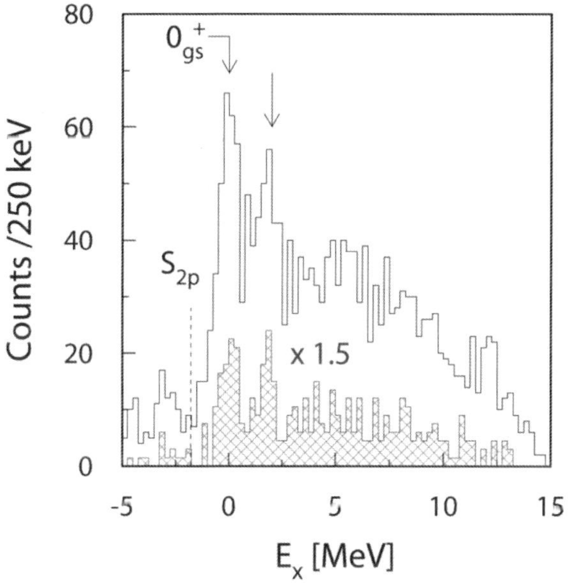

Figure 1. Reconstructed excitation energy spectrum of ^{12}O. The hatched histogram shows the spectrum gated by Θ_{cm}=35-45deg.

The excitation energy spectrum for ^{12}O is shown in Fig.1. Since ^{12}O is unbound for the ^{10}C+2p decay, the data were obtained in coincidence with the ^{14}O beam, the recoiling tritons, and the ^{10}C ejectiles. Above the ground state peak at 0 MeV, one can see another peak at around 2~MeV, indicating an excited state of ^{12}O. Besides, the spectrum exhibits a broad bump centered at 5 MeV, possible origins of which are ascribed to a superposition of resonances in ^{12}O and background from other ^{14}O reactions into the t+p+^{10}C channel. A Gaussian fit to the spectrum gives peak energies E_x of 0.0(4) and 1.8(4) MeV for the ground and excited states of ^{12}O, respectively, where the errors are dominated by the systematic error of 0.3 MeV. Differential cross sections for the ground and excited states are shown in Fig.2.

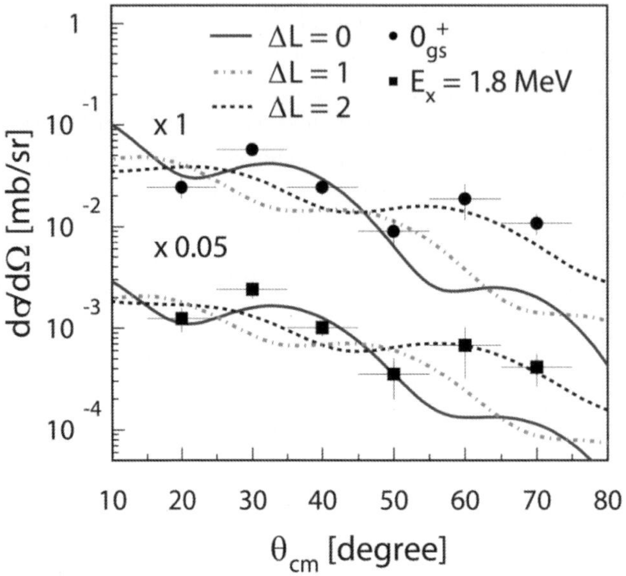

Figure 2. Differential cross-section for the $^{14}O(p,t)^{12}O$ reaction. Experimental data are compared to DWBA calculations for ΔL=0,1,2.

Vertical bars represent statistical errors only. The diffractive patterns, clearly observed for the ground and 1.8~MeV states, provide convincing evidence for both states. Data are compared with DWBA calculations performed with the code FRESCO [7] assuming a two-neutron cluster transfer. We employed global optical-model potential parameters for proton and triton. The pattern of the ΔL=1 calculation is clearly incompatible with both angular distributions. However, while the ΔL=0 distributions most closely match the data for both states, ΔL=2 cannot be completely ruled out. We therefore determine J^{π} of the newly-observed state at 1.8(4) MeV to be 0^+ or 2^+.

The 1.8 MeV state in ^{12}O illustrates an abrupt lowering of E_x, suggesting enhanced proton or neutron excitations. In ^{12}Be, the lowering of E_x of the 0^+_2 or 2^+_1 states has indicated significant neutron sd-shell intruder configurations [8]. Thus, the lowered excited state in ^{12}O strongly suggests that the Z = 8 proton shell closure is also diminishing in ^{12}O. The observed breakdown of the proton shell closure in ^{12}O implies that the underlying mechanisms for the shell evolution prevail in the vicinity of the proton drip line. The above results are detailed in [8].

2. Shell evolution around ^{68}Ni

The shell structure evolution of neutron rich nuclei around N = 40 has striking similarities with that prevailing at N = 20, which is also a magic number of "harmonic oscillator" type (as opposed to magic numbers of "spin-orbit" type) [10]. At N=20, changes in the structure is determined by the $\pi d_{5/2}$-$\nu d_{3/2}$ proton-neutron interaction. This interaction being attractive, the energy of the $\nu d_{3/2}$ orbital rises gradually as the proton layer is emptying, thus reducing the gap N = 20. Similarly, at N = 40, the gap disappears when the number of protons on the π $f_{7/2}$ shell decreases. In both cases, reducing the gap facilitates the excitation of neutrons. The quasi-degeneracy between the neutron orbitals $f_{7/2}$ and $g_{9/2}$ ($p_{3/2}$ and $d_{5/2}$ at N=20, respectively) gives rise to a sequence of spins j, j-2 beyond the Fermi surface which favors deformed configurations of 2particles-2holes type that dominate the ground-state of Cr and Fe isotopes giving rise to large deformations for these isotopes. To reproduce quantitatively the observables such as the energy of the first 2^+ state, the B (E2), and deformations in this mass region using large-scale shell model calculation, it is crucial to include orbitals such as the $2d_{5/2}$. This is also required in order to provide reliable predictions on the doubly magic character of ^{78}Ni.

In the present work, we have investigated neutron single-particle states in ^{69}Ni using the stripping reaction d(^{68}Ni,p)^{69}Ni. The experiment was performed at the GANIL facility. The ^{69}Ni projectiles were produced at an energy of 25.16 A.MeV by fragmentation of a ^{70}Zn primary beam at 62.5 A.MeV impinging on a ^{9}Be target of 500 µm thickness. The ^{68}Ni fragments were selected, using the LISE3 spectrometer with a good purity of ~90%. The reaction target was made of CD2 material of 2.6 mg/cm^2 thickness. The experimental setup placed around the target consisted of two beam tracking detectors, used to localize the reaction on the target and of four MUST2 telescopes detecting reaction light products in the backward hemisphere from between 100 and 140 degree with respect to the beam direction. For the most backward angles 150-163 degree, an annular double sided silicon strip detector was used. The detection near zero degrees of the heavy residues was performed by a plastic scintillator coupled to an ionization chamber. Finally, four EXOGAM segmented Ge detectors were placed nearly perpendicular to the beam axis at ~ 6cm from the target.

Fig. 3 presents a preliminary excitation energy spectrum reconstructed from the energy and angles of the recoiling protons. As expected the ground of ^{69}Ni is strongly populated. At around 2.5 MeV excitation energy, a prominent peak is observed that lies in the energy region where the $2d_{5/2}$ orbital is expected. Other structures are observed at higher excitation energies. Data are currently under analysis. Differential cross-sections will allow to perform L-assignents of the populated orbits and extract the related spectroscopic factors. Further details can be found in [11].

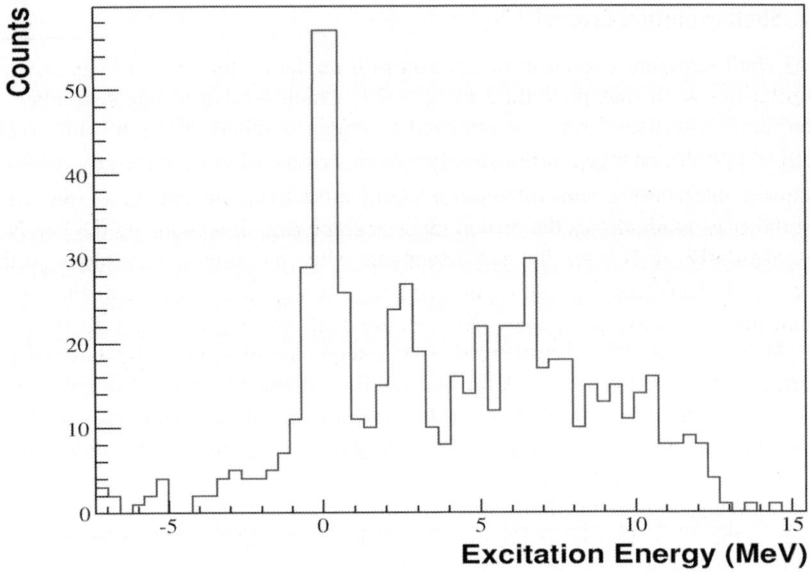

Figure 3. Reconstructed excitation energy spectrum in ^{69}Ni from the ^{68}Ni(d,p) reaction at 25.16 A.MeV.

References

1. J.M. Oliveira et al.,, *Phys. Rev. Lett.* 84, 4056 (2000).
2. K. Markenroth et al, *Phys. Rev.* **C**62, 034308 (2000).
3. A. Navin. et al., *Phys. Rev. Lett.* 85, 266 (2000).
4. H. Iwasaki et al., *Phys. Lett.* **B**491, 8 (2000).
5. S. Shimoura et al., *Phys. Lett.* **B**560, 31 (2003).
6. D. Suzuki, Ph.D thesis, University of Tokyo (2009).
7. I.J. Thompson, Comp. *Phys. Rep.* 7, 167 (1988).
8. T. Suzuki, R. Fujimoto and T. Otsuka, *Phys. Rev.* **C**67, 044302 (2003).
9. D. Suzuki et al., Phys. Rev. Lett. 103, 152503 (2009).
10. O. Sorlin et al., Eur. Phys. J. **A**16, 55 (2003).
11. M.Moukaddam et al., Acta Phys.Pol. B42, 541 (2011).

Lifetime measurement of 2_1^+ state in ^{74}Zn by recoil-distance Doppler-shift method

M. Niikura[*1], B. Mouginot[*1], F. Azaiez[*1], S. Franchoo[*1], I. Matea[*1], I. Stefan[*1], D. Verney[*1], M. Assie[*1,2], P. Bednarczyk[*3], C. Borcea[*4], A. Burger[*5], G. Burgunder[*2], A. Buta[*4], L. Cáceres[*2], E. Cléement[*2], L. Coquard[*6], G. de Angelis[*7], G. de France[*2], F. de Oliveira Santos[*2], A. Dewald[*8], A. Dijon[*2], Z. Dombradi[*9], E. Fiori[*10], C. Fransen[*8], G. Friessner[*8], L. Gaudefroy[*11], G. Georgiev[*10], S. Grévy[*2], M. Hackstein[*8], M. N. Harakeh[*2,12], F. Ibrahim[*1], O. Kamalou[*2], M. Kmiecik[*3], B. Lozeva[*10], A. Maj[*3], C. Mihai[*4], O. Möller[*6], S. Myalski[*3], F. Negoita[*4], D. Pantelica[*4], L. Perrot[*1], Th. Pissulla[*8], F. Rotaru[*4], W. Rother[*8], J. A. Scarpaci[*1], C. Stodel[*2], J. C. Thomas[*2], and P. Ujic[*2]

[*1] *IPN Orsay, CNRS/IN2P3, Université Paris-Sud 11, F-91406 Orsay Cedex, France*
[*2] *GANIL, CEA/CSM-CNRS/IN2P3, F-14076 Caen Cedex 05, France*
[*3] *IFJ PAN, PL-31342 Kraków, Poland*
[*4] *IFIN-HH, ROM-76900 Bucharest-Magurele, Romania*
[*5] *Department of Physics, University of Oslo, N-0315 Oslo, Norway*
[*6] *Institut für Kernphysik, der Technische Universität Darmstadt, D-64289 Darmstadt, Germany*
[*7] *Laboratori Nazionali di Legnaro, INFN, I-35020 Legnaro, Padova, Italy*
[*8] *Institut für Kernphysik, Universität zu Köln, D-50937 Klön, Germany*
[*9] *ATOMKI, H-4001 Debrecen, Hungary*
[*10] *CSNSM, IN2P3-CNRS, Université Paris Sud 11, F-91405 Orsay, France*
[*11] *CEA, DAM, DIF, F-91297 Arpajon, France*
[*12] *KVI, University of Groningen, NL-9747 AA Groningen, The Netherlands*

We have performed the first direct lifetime measurement of the 2_1^+ state in ^{74}Zn. The neutron-rich ^{74}Zn beam was produced by in-flight fragmentation of ^{76}Ge at the Grand Accélérateur National d'Ions Lourds and separated with the LISE spectrometer. The lifetime of the 2_1^+ state was measured by the recoil-distance Doppler-shift method with the Cologne plunger device combined with the EXOGAM detectors. The lifetime of the 2_1^+ state in ^{74}Zn was determined to be 27.0(24) ps, which corresponds to a reduced transition probability $B(E2; 2_1^+ \to 0^+) = 370(33)\ e^2\text{fm}^4$.

Keywords: lifetime measurement, exotic nuclei

1. Introduction

There has been a continuous interest in the study of the structure effects of neutron sub-shell closure at $N = 40$ close to proton magicity at $Z = 28$. The doubly-magic character of $^{68}_{28}Ni_{40}$ is originally pointed out by Bernas *et al.*[1] and investigated both from the theoretical and experimental points of view (see for instance Ref.[2] and references therein). One of the most interesting features is the sudden increase in collectivity in the vicinity of ^{68}Ni by an apparent over occupation of the neutron $1g_{9/2}$ orbit.[3,4] The role of the neutron $1g_{9/2}$ orbit in the onset of collectivity in the vicinity of ^{68}Ni has been debated in the past. It turns out that the lowest-lying three Nilsson orbitals originating from the spherical $\nu g_{9/2}$ orbital are all strongly down-sloping for both prolate and oblate shapes, which would make deformation energetically favorable in $^{72-74}$Zn. Recent shell-model calculations also pointed out the importance of the $\nu g_{9/2}$ orbit to explain the evolution of collectivity around ^{68}Ni.[5,6]

In order to investigate how collectivity develops in this region, the measurement of the reduced transition probability $B(E2)$ gives insight into its evolution. Prior to the current work, all $B(E2)$ values in the neutron-rich Ni and Zn isotopes were extracted from the Coulomb excitation cross sections.[3,4,7,8] Here, we report the first direct lifetime measurement of the 2_1^+ state in ^{74}Zn by the recoil-distance Doppler-shift method (RDDS) method[9] applied to the intermediate-energy inelastic scattering. The RDDS method allows a direct determination of the $B(E2)$ strengths, thus avoiding the complex analysis of cross sections where often many unknown matrix elements have to be fitted.

2. Experiment and result

The experiment has been performed at the Grand Accélérateur National d'Ions Lourds (GANIL). A primary beam of ^{76}Ge at an incident beam energy of 60 MeV/nucleon was used to produce a cocktail beam of ^{73,74}Zn by the projectile-fragmentation reaction on a ^{9}Be target. The cocktail beam was separated by the first half of the LISE spectrometer.[10] The secondary beam with an energy of 34 MeV/nucleon impinging on Cologne plunger device[11] consisting of a 445-μm-thick CD_2 target and a 273-μm-thick Be degrader in order to induce inelastic scattering. The outgoing particles were selected and identified by the second half of LISE. Gamma rays emitted from the reaction products were detected by 8 EXOGAM detectors,[12] which were surrounding the plunger device at angles of 45 and 135 degrees relative

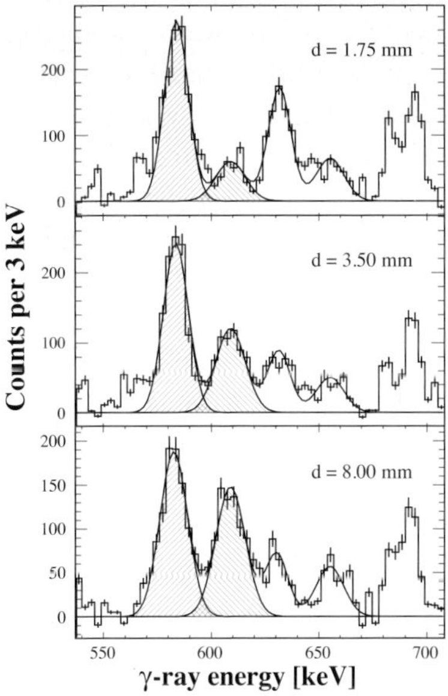

Fig. 1. γ-ray spectra after Doppler-correction and background subtraction with target-degrader distances of 1.8, 3.5 and 8.0 mm. The four photopeaks observed in the spectra are associated with decays from the 2_1^+ state in ^{72,74}Zn and the different recoil velocities before and after degrader, respectively. Solid lines represent the results of fitting.

to the beam direction. Ten different target-degrader distances of the plunger device were set over the range from 0 to 20 mm.

Figure 1 shows an example of the Doppler-corrected γ-ray spectra after background subtraction with different target-degrader distances. Solid lines represent a result of the fitting with four Gaussian functions corresponding to γ-rays emitted before and after the degrader of $2_1^+ \rightarrow 0^+$ transition in ^{72,74}Zn (653 and 606 keV), respectively. The lifetime of 2_1^+ state in ^{74}Zn was deduced based on the differential decay-curve method.[9] The decay curve was obtained from the results of fitting. We obtain the lifetime of 27.0(24) ps for the 2_1^+ state in ^{74}Zn.

3. Discussion

The obtained lifetime corresponds to $B(E2; 2_1^+ \rightarrow 0^+) = 370(33)\ e^2\text{fm}^4$, which is consistent with previously measured values from Coulomb excitation experiments at GANIL (408(30) e^2fm^4)[3] and ISOLDE (401(32) e^2fm^4).[4]In the latest work the $B(E2)$ value was obtained on the assumption of the spectroscopic quadrupole moment $Q_s = 0$ *e*b, and the authors also provide a correlation between the lifetime and Q_s (see Fig. 9 in Ref.[4]). From the comparison between the present data and the previous Coulex measurement $Q_s = +0.22(30)$ *e*b was extracted.

Spectroscopic properties of ground and excited states of even-even nuclei have been calculated in Hartree-Fock-Bogolyubov theory using the Gogny D1S interaction and the generator coordinator method.[13] These calculations predict the $B(E2; 2_1^+ \rightarrow 0^+)$ value in ^{74}Zn to be 367.6 e^2fm^4, which reproduces our result remarkably well, and they indicate triaxial deformation for ^{74}Zn with $\beta_2 = 0.299$ and $\gamma = 25°$. The experimental systematics of the excitation energy ratio $R_{42} \equiv E(4_1^+)/E(2_1^+)$ are in the range 2.0-2.3, also suggesting γ-softness in this nucleus. In such γ-soft nuclei Q_s should be close to zero, which is compatible with the extracted Q_s value within the error bar.

4. Summary

We have measured the lifetime of the 2_1^+ state in ^{74}Zn by the RDDS method at GANIL. The lifetime value was deduced to be 27.0(24) ps, consistent with previously measured values from Coulomb excitation measurements.

This work was partially supported by the German BMBF (under contract No. 06 DA 9040 I-1), the Polish Ministry of Science and Higher Education (Grant No. N N202 309135 and N N202 109936) and the Hungarian Scientific Research Fund (OTKA, contract No. K68801).

References

1. M. Bernas *et al.*, Phys. Lett. **113B**, 279 (1982).
2. K. Langanke, J. Terasaki, F. Nowacki, D. J. Dean, and W. Nazarewicz, Phys. Rev. C **67**, 044314 (2003).
3. O. Perru *et al.*, Phys. Rev. Lett. **96**, 232501 (2006).
4. J. Van de Walle *et al.*, Phys. Rev. C **79**, 014309 (2009).
5. S. M. Lenzi, F. Nowacki, A. Poves and K. Sieja, Phys. Rev. C **82**, 054301 (2010).
6. K. Sieja and F. Nowacki, Phys. Rev. C **81**, 061303(R) (2010).
7. O. Sorlin *et al.*, Phys. Rev. Lett. **88**, 092501 (2002).
8. S. Leenhardt *et al.*, Eur. Phys. J. A **14**, 1 (2002).
9. A. Dewald *et al.*, Z. Phys. A **334**, 163 (1989).
10. R. Anne *et al.*, Nucl. Instrum. Methods A **257**, 215 (1987).
11. A. Dewald *et al.*, GSI Sci. Rep. 2005, 38 (2006).
12. F. Azaiez and W. Korten, Nucl. Phys. News **7**, 21 (1997).
13. J. P. Delaroche *et al.*, Phys. Rev. C **81**, 014303 (2010).

RECENT PROGRESS OF NUCLEAR DENSITY FUNCTIONAL CALCULATIONS – TOWARD TO NEXT-GENERATION SUPERCOMPUTER –

T. INAKURA

RIKEN Nishina Center, Wako 351-0198, Japan
E-mail: inakura@riken.jp

Recently some approaches to photonuclear reaction, based on the density functional theory, have been developed. We show some results of photoabsorption cross sections for spherical and deformed nuclei. We are now ready for a systematic calculation of excited states for whole nuclear chart, using the next-generation supercomputer.

Keywords: Random phase approximation

1. Introduction

Nuclear density-functional theory (DFT) has been successful for a systematic description of ground-state properties such as binding energies, density distributions, deformation, and so on. Systematic calculations for the whole nuclear chart became feasible in the past decade.[1,2] Thanks to high performance computing, the DFT is becoming the most promising tool to quantitatively study nuclear excited state properties. The random phase approximation (RPA), which describes the nuclear excitation as a small-amplitude oscillations around the ground state, has been applied for describing both giant resonances and low-lying modes of excitations. However, it was common to approximate or ignore a part of the residual interactions and/or the residual Coulomb interaction, sacrificing the full self-consistency in practical RPA calculations. This is because the inclusion of the full residual interaction for a realistic interaction involves a cumbersome and complicated task. Moreover, almost all RPA calculations were performed in spherical nuclei in order to save the computer resources. Quite recently, some groups have reported fully self-consistent quasiparticle RPA (QRPA) calculations for deformed nuclei.[3,4] In this paper, we present recent progress of nuclear DFT calculations; the standard QRPA, the canonical-basis time-dependent Hartree-Fock-Bogoliubov (Cb-TDHFB) method, the finite amplitude method (FAM). The numerical results will be shown, mainly focused on the photoabsorption cross section.

2. Quasiparticle random-phase approximation for axially deformed nuclei

The QRPA is a standard method to calculate excited modes in heavy open shell nuclei. However, because the QRPA with realistic energy functional requires complicated programing and large computational resources, it is still a challenging subject. Recently Terasaki et al.[5] developed a paralleled computer code of the completely self-consistent QRPA based on the Hartree-Fock-Bogoliubov (HFB) states with the Skyrme functional, applicable for axially deformed nucleus. To describe the nuclear deformation and the pairing correlations, and in good account of the continuum simultaneously, they solve the HFB equations in the cylindrical coordinate space. They assume axial and parity symmetries and time-reversal invariance in the ground states. They employ the diagonalization of the traditional QRPA A, B matrix. To reduce the QRPA matrix dimension, they introduce a cut-off energy 60 MeV for the quasiparticle spectrum and truncate two-canonical-quasiparticle configurations by choosing the occupation probability $(v^{\rm pp}_{\rm cut})^2 = 10^{-6}$ and $(v^{\rm ph}_{\rm cut})^2 = 10^{-10}$. A typical number of QRPA dimension for heavy nuclei is 200,000. It takes 10,000 cpu-hours to calculate $K\pi = 1^-$ excitation mode in ^{172}Yb.

3. Canonical-basis time-dependent HFB method

Next, we present a method to calculate linear response in nuclei, based on the time-dependent DFT. We will show a feasible real-time method which is also applicable to the non-linear dynamics.

The time-dependent Hartree-Fock (TDHF) method in the three-dimensional (3D) coordinate representation is a well established method to study nuclear dynamics.[6] However, the TDHF cannot describe pairing correlations which are supposed to be important not only for static properties but also for nuclear dynamics. A straightforward extension of the TDHF including the pairing correlations is of course the TDHFB theory. However, it uses the quasiparticle orbitals whose number is, in principle, infinite. Thus, the accurate calculation of the TDHFB is now impractical. A new feasible approach we have developed is Canonical-basis TDHFB (Cb-TDHFB),[7] in which the pairing correlation is treated in the BCS level, originally proposed by Blocki and Flocard.[8]

We applied the Cb-TDHFB method to study of electric dipole resonances using the full Skyrme interaction. We calculate the time evolution of the electric dipole moment, starting from the HF+BCS ground state with an instantaneous external dipole field. Then, we perform the Fourier transform to obtain the response. The details of the calculation can be found in the reference.[7]

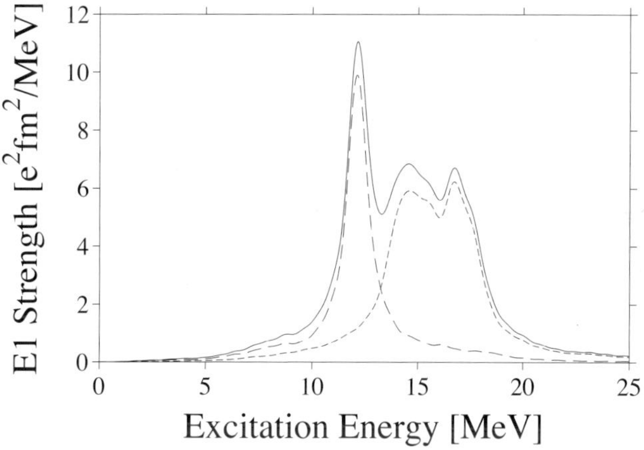

Fig. 1. Electric dipole strength distribution in deformed heavy nucleus ^{172}Yb calculated by canonical-basis TDHFB. Dashed, dotted, and solid lines correspond to $K = 0$ mode, $K = 1$ mode, and total dipole strengths, respectively. See the text for detail.

We show in Fig. 1 the calculated photoabsorption cross section in ^{172}Yb. Although the pairing potential is simplified in the Cb-TDHFB calculation, the result is quite similar to the full QRPA calculation by Terasaki et al.[5] We may conclude that the Cb-TDHFB can reproduce the QRPA result at its small amplitude limit. It should be emphasized that the computational cost of the Cb-TDHFB is significantly smaller than the QRPA. The present calculation in the full 3D space can be achieved in 300 cpu-hours, while the full QRPA calculation by Terasaki et al., which is restricted to the axially symmetric nuclei, requires 10,000 cpu-hours. This is because the Cb-TDHFB treats only the canonical states whose number is the same order as the particle number. In contrast, in the QRPA (or in the TDHFB), we need to treat the quasiparticle states whose number is the same as the dimension of the model space.

4. Finite Amplitude Method

In this section, we present the methodology of the finite amplitude method (FAM) we have developed for small-amplitude oscillation based on the time-dependent DFT.[9,10]

An explicit construction of the residual field δh requires to calculate the residual two-body kernel $v = \delta h/\delta\rho$. However, if one employs the realistic energy functional, the residual two-body kernel involves a complicated programing and a large computational cost. For this reason, it has been a common procedure to approximate or ignore a part of the residual interaction in most RPA calculations. To achieve a fully self-consistent RPA calculations, we evaluate the residual interaction δh directly from the single-particle Hamiltonian $h[\rho]$ by means of the finite deference method, thus avoiding an explicit construction of the residual two-body kernel.

The residual field δh depends linearly on the transition density. In the first order with respect to a small parameter η, we have

$$h_0 + \eta\delta h = h[\rho_0 + \eta\delta\rho] \equiv h[\tilde{\rho}_\eta] \tag{1}$$

Here, ρ_0 is the ground-state density and $h_0 = h[\rho_0]$. The one-body pseudodensity operator $\tilde{\rho}_\eta \equiv \rho 0 + \eta\delta\rho$ 1is a non-Hermitian operator expressed with bra and ket vectors as

$$\tilde{\rho}_\eta = \sum_i \left(|\phi_i\rangle + \eta|X_i\rangle\right)\left(\langle\phi_i| + \eta\langle Y_i|\right) \tag{2}$$

Here $|\phi_i\rangle$, $|X_i\rangle$, and $|Y_i\rangle$ $(i = 1, 2, \cdots, A)$ are the occupied orbitals in the ground state, forward and backward amplitudes, respectively. In the linear order in η, the residual field is expressed as

$$\delta h = \frac{1}{\eta}\left(h[\tilde{\rho}_\eta] - h_0\right) \tag{3}$$

This is the basic idea of the FAM. For more details, readers are referred to the reference.[9]

We have developed a paralleled computer program of the FAM for a Skyrme functional in the 3D coordinate-space.[10] We are undertaking a systematic calculation of the electric dipole response in even-even nuclei. So far, we calculated the photoabsorption cross section up to mass $A \sim 100$ region. Figure 2 shows a part of our achievement for nuclei in $A \sim 50$ region. The typical number of dimension of the RPA matrix in the 3D space is an order of millions for $A \sim 100$ nucleus. The FAM costs about 40 cpu-hours for each fixed excitation energy.

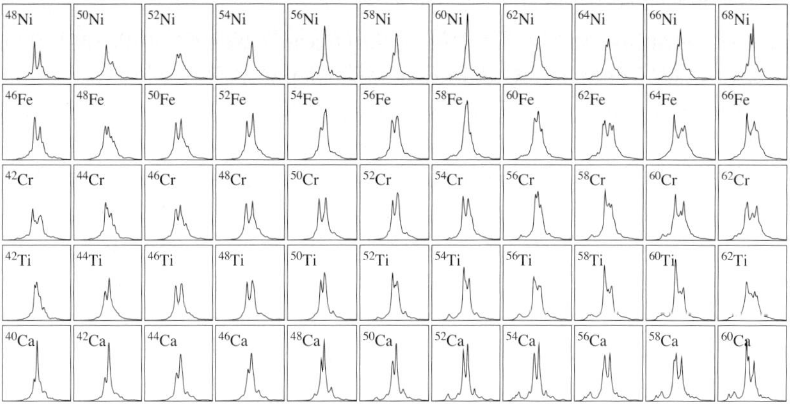

Fig. 2. Electric dipole strength distributions in mass $A \sim 50$ region. The horizontal axis corresponds to excitation energy of 0–35 MeV. The vertical axis is normalized in each isotope. The SkM* parameter set and the smoothing parameter $\Gamma = 1$ MeV are used.

The application of FAM to superfluid nuclei, namely, to the QRPA with the HFB formalism is straightforward. Quite recently, Avogadro and Nakatsukasa[11] developed the computer code of QRPA with FAM and calculated isoscalar monopole mode in spherical ^{174}Sn. They are constructing a new version of QRPA code applicable for deformed nuclei.

5. Conclusion

We have presented recent developments of density functional calculations for studies of nuclear response functions. The completely self-consistent quasiparticle random-phase approximation (QRPA) calculation has been done. It is applicable for axially deform nuclei. The paralleled QRPA code is now ready for investigation for heavy axially deform nuclei. We also present the result of the canonical-basis TDHFB. The result is almost same as the completely self-consistent QRPA result with much smaller computational cost. This is applicable to large-amplitude nuclear dynamics beyond the linear approximation. The finite amplitude method (FAM) allow us to construct the residual interaction easily and it was applied to systematic investigation of the photoabsorption cross section in light- and medium-mass region. The QRPA version of the FAM has been developed for superfluid nuclei, including the pairing correlations.

6. Acknowledgments

This work is supported by the PACS-CS (Project Nos. 09-a-25, 09-b-9, and 10a-22) of the Center for Computational Sciences, University of Tsukuba, the Large Scale Simulation Program (Project Nos. 08-14, 09-16, and 09/10-10) of High Energy Accelerator Research Organization (KEK), Grant-in-Aid for Scientific Research on Innovative Areas (Grant No. 20105003) and Grant-in-Aid for Scientific Research(B) (Grant No. 21340073). The numerical calculations were performed on PACS-CS supercomputers in University of Tsukuba, on Hitachi SR11000 at KEK, and on the RIKEN Integrated Cluster of Clusters (RICC). We also thank the International Research Network for Exotic Femto Systems (EFES) of the Core-to- Core Programs of Japan Society for the Promotion of Science.

References

1. D. Lunny, J.M. Pearson, and C. Thibault, Rev. Mod. Phys. **75**, 1021 (2003).
2. M.V. Stoitsov, J. Dobaczewski, W. Nazarewicz, S. Oittel, and D. Dean, Phys. Rev. **C 68**, 054312 (2003).
3. H. Imagawa and Y. Hashimoto, Phys. Rev. **C 67**, 037302 (2003).
4. T. Nakatsukasa and K. Yabana, Phys. Rev. **C 71**, 024301 (2005).
5. J. Terasaki and J. Engel, Phys. Rev. **C 82**, 034326 (2010).
6. J.W. Negele, Rev. Mod. Phys. **54**, 913, 1982).
7. Shuichiro Ebata, Takashi Nakatsukasa, Tsunenori Inakura, Kenichi Yoshida, Yukio Hashimoto, and Kazuhiro Yabana, Phys. Rev. **C 82**, 034306 (2010).
8. J. Blocki and H. Flocard, Nucl. Phys. **A 273**, 45 (1976).
9. Takashi Nakatsukasa, Tsunenori Inakura and Kazuhiro Yabana, Phys. Rev. **C 76**, 024318 (2007).
10. Tsunenori Inakura, Takashi Nakatsukasa, and Kazuhiro Yabana, Phys. Rev. **C 80**, 044301 (2009); Phys. Rev. **C 84**, 021302 (2011).
11. Paolo Avogadro and Takashi Nakatsukasa, Phys. Rev. **C 84**, 014314 (2011).

Tensor Correlations on Spin-Orbit Splittings and Spin Dependent Excitations

H. Sagawa

Center for Mathematics and Physics, University of Aizu, Aizu-Wakamatsu, Fukushima 965-8560, Japan
**E-mail: sagawa@u-aizu.ac.jp*

Li-Gang Cao

Institute of Modern Physics, Chinese Academy of Science, Lanzhou 730000, P.R. China

C.L. Bai

School of physics, Sichuan University, P.R. China

G. Colò

Dipartmento di Fisica, Università degli Studi and INFN, Sezione di Milano, 20133 Milano. Italy

Effects of Skyrme tensor interactions on the spin-orbit splittings are studied by using Hartree-Fock-Bogoliubov model. The charge-exchange spin-dipole (SD) excitations of ^{208}Pb are also studied by using a fully self-consistent Skyrme Hartree-Fock plus Random Phase Approximation (HF+RPA) formalism including the tensor interaction. It is found that the tensor correlations have a unique, multipole-dependent effect on the SD excitations, that is, they produce softening of 1^- states, but hardening of 0^- and 2^- states. This paves the way to a clear assessment of the strength of the tensor terms.

Keywords: 21.60.Jz, 21.65.Ef, 24.30.Cz, 24.30.Gd

1. Skyrme-Tensor Interaction

The triplet-even and triplet-odd zero-range tensor terms of the Skyrme force are expressed as[1,2]

$$V^T = \frac{T}{2}\{[(\sigma_1 \cdot \mathbf{k}')(\sigma_2 \cdot \mathbf{k}') - \frac{1}{3}(\sigma_1 \cdot \sigma_2)\,\mathbf{k}'^2]\delta(\mathbf{r}) + \delta(\mathbf{r})[(\sigma_1 \cdot \mathbf{k})(\sigma_2 \cdot \mathbf{k}) - \frac{1}{3}(\sigma_1 \cdot \sigma_2)\,\mathbf{k}^2]\}$$
$$+ \frac{U}{2}\{(\sigma_1 \cdot \mathbf{k}')\,\delta(\mathbf{r})(\sigma_2 \cdot \mathbf{k}) + (\sigma_2 \cdot \mathbf{k}')\,\delta(\mathbf{r})(\sigma_1 \cdot \mathbf{k}) - \frac{2}{3}[(\sigma_1 \cdot \sigma_2)\mathbf{k}' \cdot \delta(\mathbf{r})\mathbf{k}]\}. \quad (1)$$

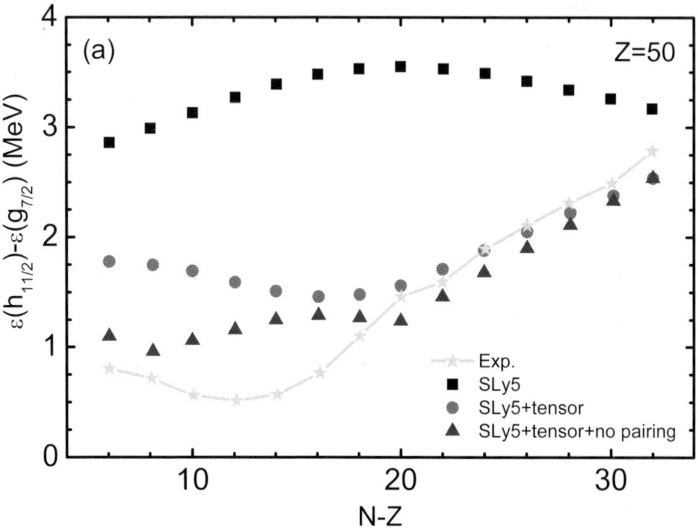

Fig. 1. (Color online) Energy gaps between $1h_{11/2}$ and $1g_{7/2}$ neutron states in Z=50 isotopes. The zero-range density dependent pairing interaction is included in the results denoted squares and circles,[3] while the pairing is not included in the results with triangles. Experimental data are taken from ref. [6]

The coupling constants T and U denote the strengths of the triplet-even and triplet-odd tensor interactions, respectively.

The main effect of the tensor terms on the HF energy density is a modification of the spin-orbit potential, which reads

$$U_{S.O.}^{(q)} = \frac{W_0}{2r}(2\frac{d\rho_q}{dr} + \frac{d\rho_{q'}}{dr}) + (\alpha\frac{J_q}{r} + \beta\frac{J_{q'}}{r}). \quad (2)$$

In this expression, q=0(1) labels neutrons (protons). ρ_q are the densities while J_q are the so-called spin-orbit densities. The first term in the r.h.s of Eq. (2) comes from the Skyrme two-body spin-orbit interaction, whereas the second term includes both a central exchange and a tensor contribution, that is, $\alpha = \alpha_C + \alpha_T$ and $\beta = \beta_C + \beta_T$: their complete expressions can be found e.g. in Ref.[3–5] In Ref.,[3] it is pointed out that both the triplet-even and triplet-odd tensor correlations play important roles to give a reasonable description of empirical energy gaps between $1h_{11/2}$ and $1g_{7/2}$ neutron orbits of Z=50 isotopes. However, other correlations such as the pairing and the particle-vibration coupling may also play important roles for quantitative discussions of single-particle energies. For example, in Fig. 1, we demonstrate the role of pairing interactions on the single-particle energies of Z=50 isotopes. We can see in the mass region N-Z=(6-14) a large pairing effect on the single-particle energies, comparing the results of the filled circles to the filled triangles.

2. Spin-Dipole Excitations and Tensor Correlations

We will show how the tensor correlations manifest themselves more decisively in the spin-dipole (SD) excitations.[7,8] The low-lying collective states are also studied in terms of the tensor correlations.[9] The charge-exchange SD operator is defined as

$$\hat{O}^{\lambda}_{\pm} = \sum_i t^i_{\pm} r_i (Y^i_1 \sigma^i)_{\lambda}. \tag{3}$$

The numerical results of the HF+RPA calculations of ^{208}Pb with the forces T43 and SLy5+T_w[7] are shown in Fig. 2. They are compared with experimental data obtained by Distorted Wave Impulse Approximation (DWIA) and multipole decomposition analysis of the (p,n) reaction.[10] From Fig.2(a) and (b) one can see that in the case of the T43 interaction the main peaks of the 0^- and 1^- strength distributions are shifted upwards by about 7.5 MeV and downwards by about 5 MeV, respectively, due to the tensor correlations. There are several 2^- peaks (cf. Fig.2(c)). The peak at excitation energy $E_x \approx 17.7$ MeV is moved upward by about 2 MeV by including tensor, and comes close to an experimental peak, while another peak at $E_x \approx 3.9$ MeV is shifted downwards by about 0.6 MeV and is also eventually close to the observed low energy peak. For the total SD strength in Fig. 2(d), it is remarkable that the main peak at 26 MeV is shifted to 21 MeV when tensor is included, and this provides good agreement with the experimental data.

In the same figure, the SD strength distributions in ^{208}Pb calculated by using the set SLy5+T_w are also shown. From Fig. 2(e), we see that the calculated 0^- strength is concentrated in one peak which is shifted upwards by about 1.3 MeV by the tensor correlations. In Fig.2(f), the RPA tensor correlations move the 1^- peak downwards and split it into three peaks, in qualitative agreement with the bump-like experimental strength. In the case of the 2^- [Fig.2(g)], the main peaks in the high energy region are rather near to the experimental main peak. Therefore, as shown in Fig.2(h), the inclusion of the tensor terms in HF+RPA can make the calculated main peak of the total SD strength coincide with the main measured peak.

We would like to obtain a better understanding of this peculiar role of tensor interactions. The diagonal matrix element of the triplet-even (TE)

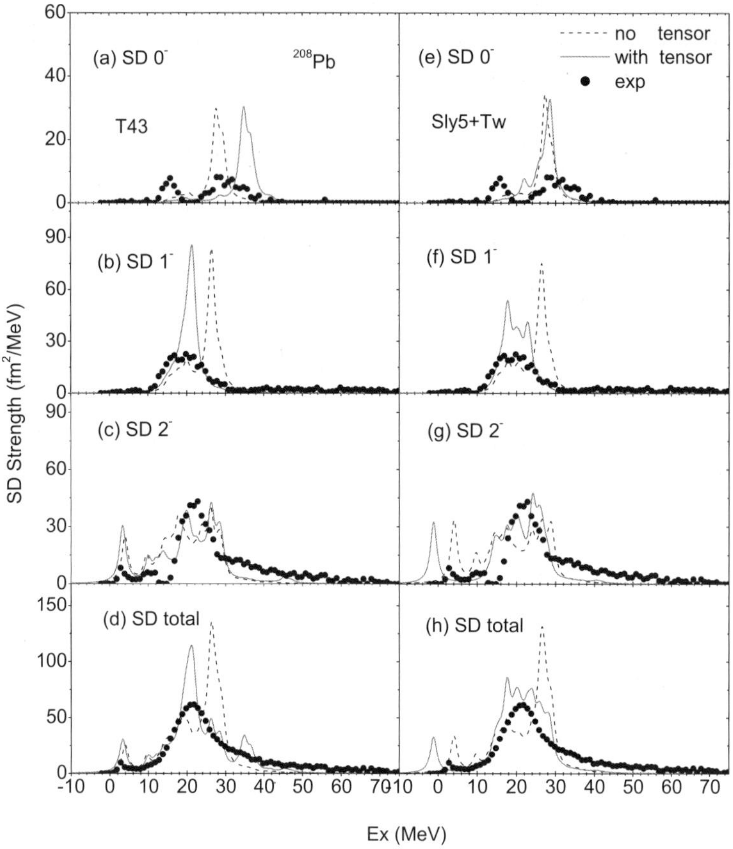

Fig. 2. (Color online) Charge-exchange SD_- strength distributions in ^{208}Pb. In the panels (a), (b), and (c) the RPA results obtained by employing the interaction SLy5+T_w for the multipoles 0^- , 1^-, 2^- are displayed. In panel (d) we show the total strength distribution. Panels (e), (f), (g) and (h) correspond to similar results when the parameter set T43 is employed. All these discrete RPA results have been smoothed by using a Lorentzian averaging with a width of 2 MeV and compared with experimental findings. The excitation energy is with respect to the ground state of ^{208}Bi. The experimental data are taken from Ref.[10]

term on a state with multipolarity λ can be expressed as[12]

$$V_{TE}^{(\lambda)} = \frac{5T}{4} \sum_{\ell,k,k'} \frac{(-)^{k+k'+\lambda+\ell+1}\hat{k}\hat{k}'}{2\lambda+1} \begin{Bmatrix} k & k' & 2 \\ 1 & 1 & \ell \end{Bmatrix} \times \begin{Bmatrix} 1 & 1 & 2 \\ k' & k & \lambda \end{Bmatrix} \langle p||\hat{O}_{k',\lambda}||h\rangle \langle p||\hat{O}_{k,\lambda}||h\rangle^*, \tag{4}$$

in terms of the reduced matrix elements of the operator $\hat{O}_{k,\lambda} = \sum_i [\sigma_i \otimes (\nabla_i \otimes Y_\ell(i))^{(k)}]^{(\lambda)}$ and $6j$ symbols. In Eq. (4), the notation $\hat{k} \equiv \sqrt{2k+1}$ is used. For the SD excitations, taking $\ell = 0$ and k=k'=1, Eq. (4) gives

$$V_{TE}^{(\lambda)} = -\frac{5}{12} T \begin{Bmatrix} 1 \\ -1/6 \\ 1/50 \end{Bmatrix} |\langle p||\hat{O}_{1,\lambda}||h\rangle|^2 \;\; for \;\; \lambda = \begin{Bmatrix} 0^- \\ 1^- \\ 2^- \end{Bmatrix}. \tag{5}$$

The triplet-odd (TO) tensor part is also expressed in a similar way as

$$V_{TO}^{(\lambda)} = \frac{5}{12} U \begin{Bmatrix} 1 \\ -1/6 \\ 1/50 \end{Bmatrix} |\langle p||\hat{O}_{1,\lambda}||h\rangle|^2 \;\; for \;\; \lambda = \begin{Bmatrix} 0^- \\ 1^- \\ 2^- \end{Bmatrix}. \tag{6}$$

We can see in Eqs. (5) and (6) that the diagonal p-h matrix element in the 0^- case is the largest, and that for 1^- is the next. The effect on 2^- is rather small. It should be noticed that these relative strengths of the Skyrme tensor interactions on each multipole are similar to those obtained from the finite-range tensor interactions both in magnitude and in sign.[11] We can sum the TE and TO direct matrix elements as

$$V_T^{(\lambda)} = V_{TE}^{(\lambda)} + V_{TO}^{(\lambda)} \equiv a_\lambda T + b_\lambda U. \tag{7}$$

The proper antisymmetrization is easy to obtain for contact interactions and gives, in the isovector channel,

$$V_{T,AS}^{(\lambda)} = [-\frac{1}{2} a_\lambda T + \frac{1}{2} b_\lambda U] \langle \tau_1 \cdot \tau_2 \rangle. \tag{8}$$

Since the coupling constant T is positive for the interactions we considered, $V_{TE}^{(\lambda)}$ is repulsive for the 0^- and 2^- case, while it is attractive for 1^-. The $V_{TO}^{(\lambda)}$ part may contribute with the same sign as the $V_{TE}^{(\lambda)}$ one if the value of U is negative. For the TIJ family, the value of U is negative or small positive, so that the $V_{TO}^{(\lambda)}$ contributions have the same multipole dependence or almost negligible. All together, the tensor correlations are strongly repulsive for 0^- and weakly repulsive for 2^- in general. For 1^-, the net effect will be attractive. For SLy5+T$_w$, the value of U is positive and will give opposite contributions to those of T. However, the T value is much larger than the value of U so that the same argument given for the TIJ family will hold.

3. Summary

In summary, we have studied the effect of tensor interactions on the charge-exchange t_- SD excitations of ^{208}Pb by using the self-consistent HF+RPA model. We have demonstrated clearly for the first time that tensor correlations have a specific multipole dependence, that is, they produce a strong hardening effect on the 0^-mode and a softening effect on the 1^- mode. A weak hardening effect is also observed on the 2^- mode. The characteristic effect of the tensor force was further examined by using the analytic formulas based on the multipole expansion of the contact tensor interaction. It is found that both the TE and TO tensor interactions do provide the characteristic multipole dependent effect on the SD excitations. These effects are shown to be robust irrespective of the adopted Skyrme force with tensor interactions. Our calculated results are compared with recent SD excitation spectra obtained in the (p, n) reaction on the ^{208}Pb target. The softening and the hardening effects produced by tensor correlations on the 1^- and 2^- modes are confirmed by comparing the experimental data and the calculated results with and without tensor correlations. Consequently, from the study of SD excitations, we are able to give a clear constraint on the effective tensor force which was missing so far.

Acknowledgments

L.G. Cao acknowledges the support of the National Science Foundation of China under Grant Nos. 10875150. This work is supported in part by the Japanese Ministry of Education, Culture, Sports, Science and Technology by Grant-in-Aid for Scientific Research under the program number (C(2))20540277.

References

1. T. H. R. Skyrme, Nucl. Phys. 9 (1959) 615.
2. Fl. Stancu, D. M. Brink and H. Flocard, Phys. Lett. 68B (1977) 108.
3. G. Colò, H. Sagawa, S. Fracasso, P.F. Bortignon, Phys. Lett. **B646**, 227 (2007).
4. T. Lesinski, M. Bender, K. Bennaceur, T. Duguet, and J. Meyer, Phys. Rev. **C76**,014312 (2007).
5. W. Zou *et al.*, Phys. Rev. C77, 014304 (2008).
6. J. P. Schiffer et al., Phys. Rev. Lett. **92** (2004) 162501.
7. C.L. Bai, H.Q. Zhang, H. Sagawa, X.Z. Zhang, G. Colò, and F. R. Xu, Phys. Rev. Lett 105, 072501(2010).
8. C.L. Bai, H.Q. Zhang, H. Sagawa, X.Z. Zhang, G. Colò, and F. R. Xu, Phys. Rev. **C83**, 054316(2011).
9. Li-Gang Cao, H. Sagawa and G. Colò, Phys. Rev. C83, 034324(2011).
10. T. Wakasa, arXiv:1004.5220[nucl-ex].
11. D. M. Brink and Fl. Stancu, Phys. Rev. C75, 064311 (2007).
12. L. Cao *et al.*, Phys. Rev. C80, 064304 (2009).

Systematic study of many-particle and many-hole states in and around the "island of inversion"

M. Kimura

Creative Research Institution (CRIS), Hokkaido University, N21W10, Sapporo 001 0021, Japan.

Deformation and many particle and many hole configurations in the island of inversion are investigated with antisymmetrized molecular dynamics (AMD). The result suggests the coexistence of many particle and many hole configurations with different deformation at small excitation energy. Detailed study of the excitation spectrum is performed for ^{33}Mg. We suggest the negative-parity assignment for the ground state, and the coexistence of many particle hole states at very small excitation energy is predicted.

1. Introduction

The breaking of neutron magic number $N = 20$ has been one of the great interests in nuclear physics. One of the recent experimental finding is that there are coexistence of "spherical and deformed shape" or "normal and intruder configurations" in many cases.[7–10] It is expected that the many particle hole states with different deformation coexist at very small excitation energy (shape coexistence), since the quenching of the $N = 20$ shell gap may induces strong quadrupole correlations in the island of inversion.

We report our recent study of the shape coexistence in the island of inversion based on antisymmetrized molecular dynamics (AMD). AMD has been successfully applied and reproduced properties of nuclei in and around the island of inversion. We first demonstrate the onset of deformation in the island of inversion by investigating the energy curve as function of quadrupole deformation. Then we discuss the detailed excitation spectrum of ^{33}Mg. Since the spectroscopy of odd-neutron nuclei is directly related to the last neutron orbit and deformation of the system, ^{33}Mg is a good probe to detect the shape coexistence. It is shown that the ground state of ^{33}Mg is well deformed negative-parity state with a $3p2h$ configuration, that nicely explains the observed data. Coexistence of spherical and more deformed states with different particle hole configuration at very small excitation energy is predicted.

2. Framework of AMD

For the detail of AMD framework, readers are directed to the references.[13,14] The Hamiltonian used in this study is given as,

$$H = \sum_{i}^{A} t_i + \sum_{i<j}^{A} v_N(ij) + \sum_{i<j\in p} v_C(ij) - T_{c.m.}. \tag{1}$$

As an effective nuclear force v_N, we have used Gogny D1S force, and the Coulomb force is approximated by the sum of seven Gaussians. $T_{c.m.}$ takes care of the center-of-mass energy.

We employ the parity projected Slater determinant as the variational wave function,

$$\Phi^{\pi} = \hat{P}^{\pi}\Phi_{\text{int}} = \frac{(1+\pi\hat{P}_x)}{2}\Phi_{\text{int}}, \tag{2}$$

$$\Phi_{\text{int}} = \mathcal{A}\{\varphi_1, \varphi_2, ..., \varphi_A\}, \quad \varphi_i(\mathbf{r}) = \phi_i(\mathbf{r})\chi_i\xi_i, \tag{3}$$

and the intrinsic wave function is represented by a Slater determinant, where φ_i is the ith single particle wave packet consisting of the spatial ϕ_i, spin χ_i and isospin ξ_i parts. ϕ_i is described by a triaxially deformed Gaussian,

$$\phi_i(\mathbf{r}) = \exp\left\{-\sum_{\sigma=x,y,z} \nu_\sigma \left(r_\sigma - \frac{Z_{i\sigma}}{\sqrt{\nu_\sigma}}\right)^2\right\}, \quad \chi_i = \alpha_i\chi_\uparrow + \beta_i\chi_\downarrow, \quad \xi_i = \text{p or n} \tag{4}$$

The centroids of wave packets ($\boldsymbol{Z}_i$) and direction of spins (α_i, β_i) are optimized to minimize the total energy under the constraint on the matter quadrupole deformation β. After the variation, we project out the eigenstate of the total angular momentum, $\Phi^{J\pi}_{MK}(\beta) = \hat{P}^J_{MK}\Phi^{\pi}(\beta)$. Finally, the wave functions $\Phi^{J\pi}_{MK}(\beta)$ which have the same parity and angular momentum but have different K and deformation β are superposed (GCM). Then the wave function of the system is written as

$$\Psi^{J\pi}_{\alpha} = \sum_{Ki} c_{Ki,\alpha}\Phi^{J\pi}_{MK}(\beta_i) \tag{5}$$

The coefficients c_n, c'_n,... are determined by the Hill-Wheeler equation.

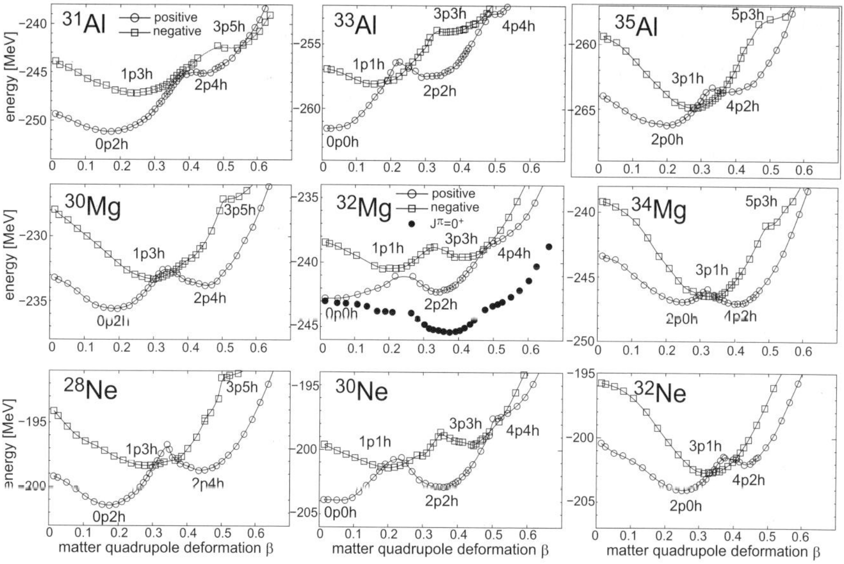

Fig. 1. Energy curves of Ne, Mg and Al isotopes as function of quadrupole deformation. Open circles (open boxes) show the positive-parity (negative-parity) states before the angular momentum projection for ^{32}Mg.

3. Onset of the deformation in the island of inversion

Fig. 1 shows the energy curves of Ne, Mg and Al isotopes with $N = 18$, 20 and 22 calculated with AMD. Here Ne and Mg isotopes with $N = 20$ and 22 are in the island and Al isotopes are out of or on the border. In the case of ^{32}Mg, there are three minima in the positive-parity energy curve and two in the negative-parity. These minima have different single-particle configurations. It is confirmed by the neutron single particle energies as function of quadrupole deformation. In the spherical region, the last two neutrons occupy the orbital originates in *sd*-shell, and the system has a $0p0h$ configuration. Around $\beta = 0.3$ two neutrons occupy an orbital intruding from *pf*-shell and the system has a $2p2h$ configuration. Further increase of deformation causes the intrusion of another orbital from *pf*-shell ($4p4h$ configuration). Thus, the minima on the positive-parity energy curve are understood to have $0p0h$, $2p2h$ and $4p4h$ configurations in ascending order of deformation. In the same way, the minima on the negative-parity energy curve have $1p1h$ and $3p3h$ configurations.

From Fig. 1, we can see that different particle-hole configurations with different deformation always exist even in nuclei out of the island of inversion. When we go into the island of inversion, we observe that the minima with many particle-hole configurations with large deformation come down and coexist with spherical states.

4. Low-lying level structure of ^{33}Mg

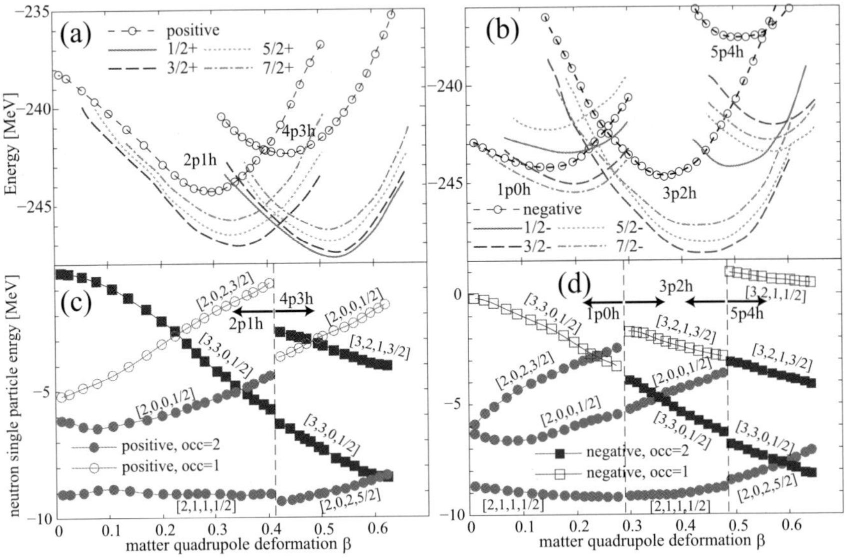

Fig. 2. (color online) (a) and (b): Energy curves of ^{33}Mg before and after angular momentum projection for (a) positive- and (b) negative-parity states. Circles with dashed lines show the energy curve before the angular momentum projection. Colored lines show the energy curves after the angular momentum projection to $J = 1/2$, $3/2$, $5/2$ and $7/2$. (c) and (d): Corresponding single particle energies of the last 7 neutrons. Filled (open) symbols show the orbits occupied by two (one) neutrons. Circles (boxes) show the orbits in which positive-parity (negative-parity) component is larger than 50 %.

The system with odd number neutrons may be the best example to show the coexistence of shapes and *mpmh* configurations. Fig. 2 shows the calculated energy curve of positive- and negative-parity states before and after the angular momentum projection. It shows that the strongly deformed $3p2h$ ($J^\pi = 3/2^-$) configuration has the lowest energy, and other particle configurations from $1p0h$ to $5p4h$ coexist below 5 MeV in excitation energy. It is notable that these particle-hole configurations have different spin-parity as the lowest state. In the bottom of the Fig. 2, the neutron single particle levels are shown. We can confirm from it that the spin-parity

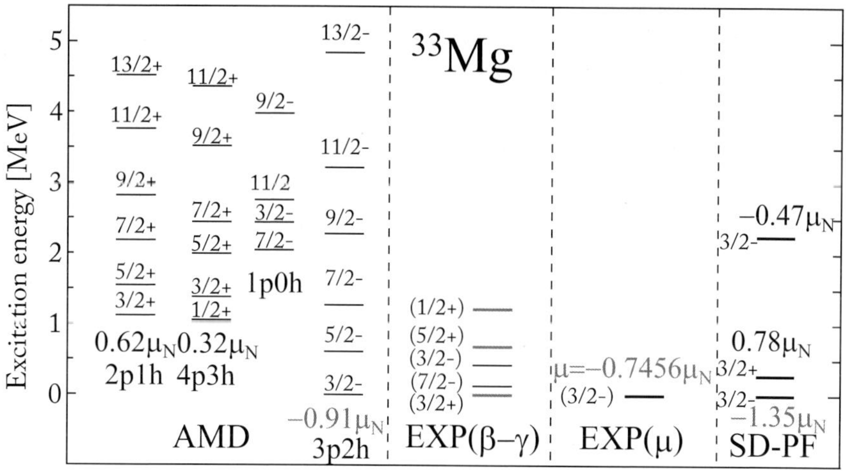

Fig. 3. Calculated spectrum of ^{33}Mg compared with the tentative assignments suggested in Refs.[16,17] Numbers show excitation energies in MeV.

of the lowest state of each configuration is determined by the property of the Nilsson-like orbit the last neutron occupies. Therefore, by looking the spectrum of the odd neutron system, we can investigate the shape coexistence and neutron orbital in the island of inversion.

Fig. 3 shows the calculated spectrum and the magnetic moment of ^{33}Mg, and compares them with observations and the shell model calculation. There are conflicting assignments for the ground state spin-parity based on several experiments. From the observation of β decay from ^{33}Na[15] and to ^{33}Al[16] the positive-parity assignment is suggested, while the negative-parity is suggested from the measurement of the magnetic moment[17] and $1n$ removal cross section.[18] Looking at the AMD result, we can see that $1p0h$, $2p1h$, $3p2h$ and $4p3h$ configurations coexist within very small excitation energy. The spin parity of these band head states are basically consistent with the last neutron single particle orbitals. It's behavior is quite similar to that of ^{32}Mg shown in Fig.2. Namely, at small deformation, the last neutron occupies [3,3,0,1/2] orbital ($1p0h$) with $J^{\pi} = 7/2^-$. As deformation becomes larger, the last neutron occupies [2,0,2,3/2] ($2p1h$), [3,2,1,3/2] ($3p2h$) and [2,0,0,1/2] ($4p3h$) orbitals that respectively generate $K^{\pi} = 3/2^+, 3/2^-$ and $1/2^+$ bands. Among these bands, only the J^{π}=$3/2^-$ state has the negative-sign of the magnetic moment and consistent with the observation of the magnetic moment. Note that AMD calculation predicts the *mpmh* configurations with different deformation at small excitation energy and further experimental studies are needed.

5. Summary

We have investigated deformation and *mpmh* configurations in the island of inversion based on AMD. The energy curves show the coexistence of *mpmh* configurations with different deformation. In the case of ^{33}Mg, these *mpmh* configurations coexist at very small excitation energy. AMD calculation supports the negative-parity assignment for the ground state suggested by the observation of the magnetic moment. The calculation also predicts different *mpmh* configurations and further experimental investigations are requested.

References

1. C. Thibault *et al.*, Phys. Rev. **C 12**, 644 (1975).
2. G. Huber *et al.*, Phys. Rev. C **18**, 2342 (1978).
3. A. Poves and J. Retamosa, PHys. Lett. B **184**, 311 (1986).
4. E. K. Warburton, J. A. Becker and B. A. Brown, Phys. Rev. C **41**, 1147 (1990).
5. N. Fukunishi, T. Otsuka and T. Sebe, Phys. Lett. B **296**, 279 (1992).
6. T. Motobayashi *et al.*, Phys. Lett. B **346**, 9 (1995).
7. S. Takeuchi *et al.*, Phys. Rev. C **79**, 054319 (2009).
8. W. Schwerdtfeger *et al.*, Phys. Rev. Lett. **103**, 012501 (2009).
9. Z. M. Wang *et al.*, Phys. Rev. C **81**, 064301 (2010).
10. K. Wimmer *et al.*, Phys. Rev. Lett. **105**, 252501 (2010).
11. M. Kimura, Phys. Rev. C **75**, 041302 (2007).
12. D. Miller *et al.*, Phys. Rev. C **79**, 054306 (2009).
13. Y. Kanada-En'yo, H. Horiuchi and A. Ono, PHys. Rev. **C52**, 628 (1995).
14. M .Kimura, Phys. Rev. **C69**, 044319 (2004).
15. S. Nummela, *et. al.*, Phys. Rev. **C64**, 054313 (2001).
16. V. Tripathi, *et. al.*, Phys. Rev. Lett. *101*, 142504 (2008).
17. D. T. Yordarov, *et. al.*, Phys. Rev. Lett. **99**, 212501 (2007).
18. R. Kanungo, *et. al.*, Phys. Lett. **B685**, 253 (2010).

Coexistence of various deformed states and α clustering in ^{42}Ca

Yasutaka TANIGUCHI

RIKEN Nishina Center for Accelerator-Based Science, RIKEN, Wako, Saitama 351–0198, Japan
**E-mail: yasutaku@riken.jp*

Structures in low-lying states of ^{42}Ca have been studied focusing on various deformations and α cluster correlations using the antisymmetrized molecular dynamics and the generator coordinate method. Three deformed states coexist in low-lying states of ^{42}Ca, and one of them contain much α–^{38}Ar cluster structure components.

Keywords: Coexistence of various deformations; α clustering

1. Introduction

In this ten years, various deformed states in low-lying states in $A = 40$ region have been confirmed.[1–5] To study universal existence of deformed states in this region, structures in low-lying states of ^{42}Ca have been studied.

Excited states in ^{42}Ca was studied by a γ-spectroscopy experiment.[4] It observed a rotational band built on the $J^\pi = 0_2^+$ state up to (12^+) state. Particle-hole configurations of low-lying states were studied by an α transfer experiment.[6] α transfer reactions populate the $J^\pi = 0_1^+$ and 0_3^+ states strongly and the $J^\pi = 0_2^+$ state weakly. It concluded that the $J^\pi = 0_2^+$ and 0_3^+ states have $6p4h$ and $4p2h$ configurations, respectively, relative to the ground state of ^{40}Ca.

Theoretically, an α + ^{38}Ar orthogonal condition model shows that the rotational band built on the $J^\pi = 0_2^+$ state have α–^{38}Ar cluster structure and main components of the states are $4p2h$ configurations, which is inconsistent with experimental results.

2. Framework

In this study, degrees of freedoms of quadrupole deformation β and inter-cluster motions between α and ^{38}Ar clusters are treated by using the gener-

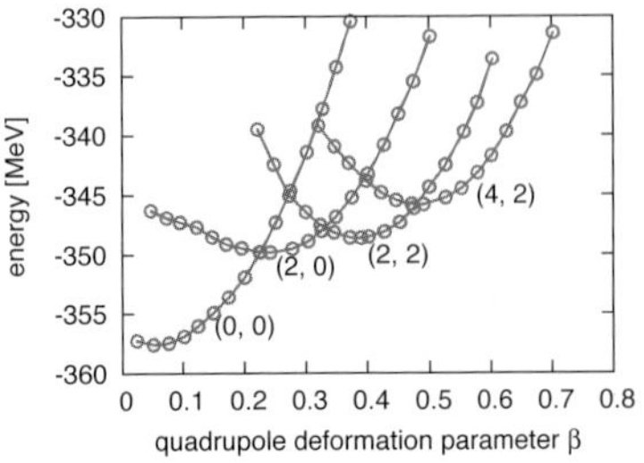

Fig. 1. Binding energies as functions of quadrupole deformation parameter β obtained by energy variations with β constraint are shown. See text about numbers in parentheses.

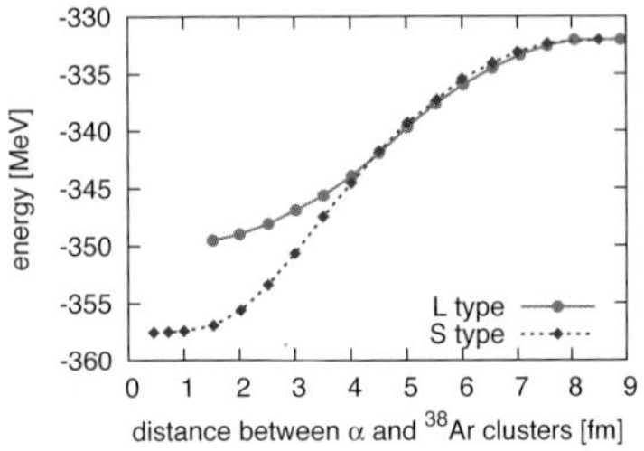

Fig. 2. Energies of α–^{38}Ar cluster structures as functions of inter-cluster distances obtained by inter-cluster distance constrained energy variations. Red and blue lines are for L and S type, respectively.

ator coordinate method (GCM), which is superposition of basis wave functions to diagonalize the Hamiltonian. Trial wave functions are the deformed-basis antisymmetrized molecular dynamics (AMD) wave functions, which is the Slater determinant of Gaussian wave packets that can deform triaxially.

GCM basis are obtained by energy variation with constraints on β and inter-cluster distance between α and ^{38}Ar clusters. Harmonic oscillator quanta constraints for protons and neutrons are also constrained in the β constrained energy variations. The GCM basis are projected to eigenstates of parity and angular momentum before and after energy variation, respectively. The Gogny D1S force was used.

3. Results

Figure 1 shows β-energy curves obtained by energy variations with β constraint. Numbers in parentheses show harmonic oscillator quanta of protons and neutrons for left and right numbers, respectively, relative to the ground state. On the energy surface, (0, 0), (2, 0), (2, 2) and (4, 2) configurations are obtained from small to large β.

Figure 2 shows energies of α–^{38}Ar cluster structures as functions of inter-cluster distances obtained by inter-cluster distance constrained energy variations. An L and S types show α clusters locate on long and short axises of slightly deformed ^{38}Ar clusters, respectively. In small distance region, energies of the S type are close to the ground state energy. On the other hand, the L type is still excited.

Figure 3 shows harmonic oscillator quanta for α–^{38}Ar cluster structures

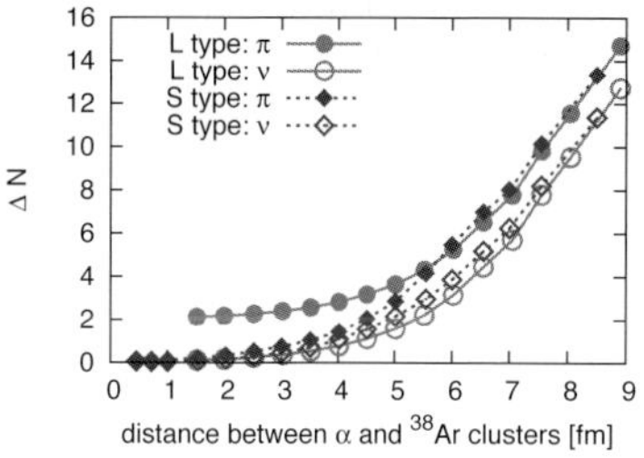

Fig. 3. Harmonic oscillator quanta for α–^{38}Ar cluster structures as functions of inter-cluster distances are shown. Red and blue lines show the L and S types, respectively, and open and close symbols shows protons and neutrons, respectively.

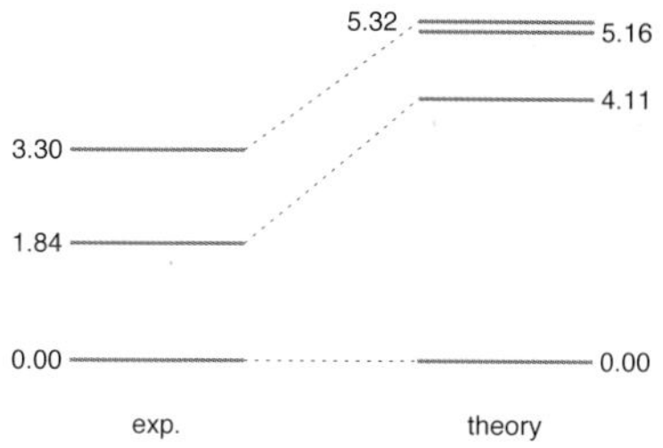

Fig. 4. Levels of $J^\pi = 0^+$ states are shown. Left and right parts shows experimental and theoretical values, respectively.

as functions of inter-cluster distances. In small distance region, S type wave functions become (0, 0) configurations, which means those wave functions are close to the ground state. On the other hand, L type wave functions become (2, 0) configurations.

Figure 4 shows results of the GCM for $J^\pi = 0^+$ states. Dominant components of them are $2p$, $6p4h$, $8p6h$ and $4p2h$ states obtained by β constrained energy variations from the $J^\pi = 0_1^+$ to 0_4^+ states, respectively. Their particle-hole configurations shows that the theoretical $J^\pi = 0_2^+$ and 0_4^+ states correspond to the experimental $J^\pi = 0_2^+$ and 0_3^+, respectively. The $J^\pi = 0_2^+$, 0_3^+ and 0_4^+ states are named as ND1, SD and ND2 states, respectively, for β values of those dominant components. The SD state with $8p6h$ configuration has never been identified.

The $J^\pi = 0_1^+$ and 0_4^+ states contain S type and L α–^{38}Ar cluster structure components, respectively.

4. Discussions and conclusions

ND1, ND2 and SD states in ^{42}Ca have (2, 0), (2, 2) and (4, 2) configurations, respectively. Each particle-hole configuration for protons and neutrons favor deformation. For proton part, $2\hbar\omega$ excited configurations, $(sd)^{-2}(pf)^2$, and $4\hbar\omega$ excited configurations, $(sd)^{-4}(pf)^4$, correspond to normal-deformed and superdeformed (SD) states, respectively, in ^{42}Ca ($N = Z = 20$). For neutron part, $2\hbar\omega$ excited configurations, $(sd)^{-2}(pf)^4$, correspond to SD states in ^{44}Ti ($N = Z = 22$). Particle-hole configurations that protons and neutrons favor in deformed states and its combinations develop rich structures in ^{42}Ca.

The $J^{\pi} = 0_1^+$ and 0_4^+ states contain much α–^{38}Ar cluster structure components, which is consistent with the α transfer experiment. To explain these populations by α transfer reactions, an α–^{38}Ar cluster structure picture is quite powerful. The $J^{\pi} = 0_1^+$ and 0_4^+ states contain S and L type components, respectively. They have (0, 0) and (2, 0) configurations, respectively, in small inter-cluster distance region. Those are same configurations as dominant components of the $J^{\pi} = 0_1^+$ and 0_3^+ states, respectively. Particle-hole configurations of cluster structures in small inter-cluster distance are important for their coupling with low-lying deformed states.

References

1. C. D. O'Leary, M. A. Bentley, B. A. Brown, D. E. Appelbe, R. A. Bark, D. M. Cullen, S. Ertürk, A. Maj and A. C. Merchant, *Phys. Rev. C* **61**, 064314 (2000).
2. E. Ideguchi, D. G. Sarantites, W. Reviol, A. V. Afanasjev, M. Devlin, C. Baktash, R. V. F. Janssens, D. Rudolph, A. Axelsson, M. P. Carpenter, A. Galindo-Uribarri, D. R. LaFosse, T. Lauritsen, F. Lerma, C. J. Lister, P. Reiter, D. Seweryniak, M. Weiszflog and J. N. Wilson, *Phys. Rev. Lett.* **87**, 222501 (2001).
3. C. E. Svensson, A. O. Macchiavelli, A. Juodagalvis, A. Poves, I. Ragnarsson, S. Åberg, D. E. Appelbe, R. A. E. Austin, G. C. Ball, M. P. Carpenter, E. Caurier, R. M. Clark, M. Cromaz, M. A. Deleplanque, R. M. Diamond, P. Fallon, R. V. F. Janssens, G. J. Lane, I. Y. Lee, F. Nowacki, D. G. Sarantites, F. S. Stephens, K. Vetter and D. Ward, *Phys. Rev. C* **63**, 061301 (2001).
4. M. Lach, J. Styczen, W. Meczynski, P. Bednarczyk, A. Bracco, J. Grebosz, A. Maj, J. C. Merdinger, N. Schulz, M. B. Smith, K. M. Spohr, J. P. Vivien and M. Zieblinski, *Eur.Phys.J. A* **16**, 309 (2003).
5. E. Ideguchi, S. Ota, T. Morikawa, M. Oshima, M. Koizumi, Y. Toh, A. Kimura, H. Harada, K. Furutaka, S. Nakamura, F. Kitatani, Y. Hatsukawa, T. Shizuma, M. Sugawara, H. Miyatake, Y. Watanabe, Y. Hirayama and M. Oi, *Phys. Lett. B* **686**, 18 (2010).
6. H. Fortune, R. Betts, J. Bishop, M. AL-Jadir and R. Middleton, *Nucl. Phys. A* **294**, 208 (1978).

Tops-on-top model for triaxial strongly deformed bands in even-A nuclei

Kazuko Sugawara-Tanabe* and Kosai Tanabe

Theoretical Nuclear Physics Laboratory, RIKEN, Wako, Saitama 351-0198, Japan
**E-mail: s-tanabe@riken.jp*

We extend top-on-top model to even-A nuclei, where two particles in different single-particle orbitals (tops) correlate with the triaxial rotor (top). We apply the model to describe the energy levels and transition rates for the triaxial, strongly deformed (TSD) bands in odd-odd nucleus ^{164}Lu.

Keywords: TSD bands, tops-on-top model, Coriolis coupling

1. Formalism

We have obtained an algebraic solution for a top-on-top model, where one high-j nucleon is coupled to a triaxially deformed rotor,[1–3] by introducing two kinds of bosons for the total angular momentum $\vec{I}$ and the single-particle angular momentum $\vec{j}$ with the use of Holstein-Primakoff (HP) transformation. The main characteristics of top-on-top model are as follows: (1) Top-on-top model deals with the Coriolis coupling between $\vec{j}$ and $\vec{I}$ explicitly. (2) An approximation is attained by taking account of the next-to-leading order in HP boson expansion in a way consistent with the D_2 invariance. (3) In order to simulate the experimental data, we have introduced the angular momentum-dependence of the rigid moments of inerta with two parameters c_1 and c_2:

$$\mathcal{J}_k = \frac{\mathcal{J}_0^{\text{eff}}(I)}{1+(\frac{5}{16\pi})^{1/2}\beta_2}\left[1-\left(\frac{5}{4\pi}\right)^{1/2}\beta_2\cos\left(\gamma+\frac{2}{3}\pi k\right)\right];\ \mathcal{J}_0^{\text{eff}}(I) = \mathcal{J}_0\frac{I-c_1}{I+c_2}, \tag{1}$$

where $k = 1, 2$ and 3 correspond to x, y and z, and β_2 and γ are the deformation parameters. Our formula is applicable to the high spin and highly excited region, i.e., TSD band region, so that $\mathcal{J}_0^{\text{eff}}(I)$ increases monotonically, and goes to $\mathcal{J}_0$ as $I \to \infty$. It is a sort of Padé approximation for the polynomial in $1/I$, and simulates the Coriolis anti-pairing effect in the core moments of inertia. As this angular momentum dependence has quite well reproduced the excitation energy relative to the reference, i.e. $E^* - aI(I+1)$ in odd-A Lu isotopes, we extend this top-on-top model to odd-odd nucleus.

We start from the Hamiltonian with two non-interacting particles outside the triaxial core,

$$H = H_{\rm rot} + H_{\rm sp}(\vec{j}_1) + H_{\rm sp}(\vec{j}_2) \tag{2}$$

with

$$\begin{aligned} H_{\rm rot} &= \sum_{k=x,y,z} A_k (I_k - j_{1k} - j_{2k})^2, \\ H_{\rm sp}(\vec{j}) &= \frac{V_j}{j(j+1)} \left[\cos\gamma (3j_z^2 - \vec{j}^2) - \sqrt{3} \sin\gamma (j_x^2 - j_y^2) \right], \end{aligned} \tag{3}$$

where $A_k = 1/(2\mathcal{J}_k)$, and $\vec{j}$ stands for either $\vec{j}_1$ or $\vec{j}_2$. The maximum moment of inertia is about x-axis and the relation $\mathcal{J}_x \geq \mathcal{J}_y \geq \mathcal{J}_z$ holds in the range of $0 \leq \gamma \leq 2\pi/3$. We choose x-axis as a quantization axis, and assume that two particles occupy different orbitals or different isospins, so that $[j_{1k}, j_{2k'}] = 0$ for k or $k' = x, y, z$. Then, three angular momenta $\vec{I}$, $\vec{j}_1$ and $\vec{j}_2$ are realized in terms of three kinds of Holstein-Primakoff (HP) boson operators $\hat{a}$, $\hat{a}^\dagger$, $\hat{b}$, $\hat{b}^\dagger$ and $\hat{c}$, $\hat{c}^\dagger$ as follows:

$$\begin{aligned} I_+ &= I_-^\dagger = I_y + iI_z = -\hat{a}^\dagger \sqrt{2I - \hat{n}_a}, \quad I_x = I - \hat{n}_a, \\ j_{1+} &= j_{1-}^\dagger = j_{1y} + ij_{1z} = \sqrt{2j_1 - \hat{n}_b}\, \hat{b}, \quad j_{1x} = j_1 - \hat{n}_b, \\ j_{2+} &= j_{2-}^\dagger = j_{2y} + ij_{2z} = \sqrt{2j_2 - \hat{n}_c}\, \hat{c}, \quad j_{2x} = j_2 - \hat{n}_c, \end{aligned} \tag{4}$$

where $\hat{n}_a = \hat{a}^\dagger \hat{a}$, $\hat{n}_b = \hat{b}^\dagger \hat{b}$ and $\hat{n}_c = \hat{c}^\dagger \hat{c}$. Here, we remark the commutation relation of $[I_k, I_{k'}] = -iI_{k\times k'}$, and $[j_k, j_{k'}] = ij_{k\times k'}$ for both of $\vec{j}_1$ and $\vec{j}_2$. In applying these HP transformations to Eq. (2), we expand $\sqrt{2I - \hat{n}_a}$, $\sqrt{2j_1 - \hat{n}_b}$ and $\sqrt{2j_2 - \hat{n}_c}$ into series in $\hat{n}_a/(2I)$, $\hat{n}_b/(2j_1)$ and $\hat{n}_c/(2j_2)$, and retain up to the next-to-leading order. For the case of $V_1 = V_2 = 0$, diagonalization of $H_{\rm rot}$ is achieved by the unitary transformation to quasibosons:

$$\begin{pmatrix} \alpha \\ \beta \\ \gamma \end{pmatrix} = K \begin{pmatrix} \hat{a} \\ \hat{b} \\ \hat{c} \end{pmatrix} + M \begin{pmatrix} \hat{a}^\dagger \\ \hat{b}^\dagger \\ \hat{c}^\dagger \end{pmatrix}, \tag{5}$$

where two 3×3 matrices are given by

$$K = \begin{pmatrix} \left(\frac{I}{I-j_1-j_2}\right)^{1/2}\eta_+ & -\left(\frac{j_1}{I-j_1-j_2}\right)^{1/2}\eta_- & -\left(\frac{j_2}{I-j_1-j_2}\right)^{1/2}\eta_- \\ 0 & \left(\frac{Ij_1}{(I-j_1-j_2)(j_1+j_2)}\right)^{1/2} & \left(\frac{Ij_2}{(I-j_1-j_2)(j_1+j_2)}\right)^{1/2} \\ 0 & -\left(\frac{j_2}{j_1+j_2}\right)^{1/2} & \left(\frac{j_1}{j_1+j_2}\right)^{1/2} \end{pmatrix}, \tag{6}$$

$$M = \begin{pmatrix} -\left(\frac{I}{I-j_1-j_2}\right)^{1/2}\eta_- & \left(\frac{j_1}{I-j_1-j_2}\right)^{1/2}\eta_+ & \left(\frac{j_2}{I-j_1-j_2}\right)^{1/2}\eta_+ \\ \left(\frac{j_1+j_2}{I-j_1-j_2}\right)^{1/2} & 0 & 0 \\ 0 & 0 & 0 \end{pmatrix} \tag{7}$$

with

$$\eta_\pm = \left\{ \begin{matrix} 1 \\ \mathrm{sgn}(q-p) \end{matrix} \right\} \left[\frac{1}{2}\left(\frac{p+q}{2\sqrt{pq}} \pm 1 \right) \right]^{1/2}, \tag{8}$$

where $p \equiv A_y - A_x$, $q \equiv A_z - A_x$. The eigenvalue of $H_{\rm rot}$ becomes

$$E_{\rm rot}(I, n_\alpha, n_\beta, n_\gamma) = A_x R(R+1) - \frac{p+q}{2} n_\alpha^2 + \left(2R\sqrt{pq} + \sqrt{pq} - \frac{p+q}{2} \right) \left(n_\alpha + \frac{1}{2} \right), \tag{9}$$

with $R = I - j_1 - j_2 + n_\beta + n_\gamma$. In the symmetric limit of $A_y = A_z$, $E_{\rm rot}$ takes the well-known form of $A_z R(R+1) + (A_x - A_z)(R - n_\alpha)^2$. Therefore n_β and n_γ describe precession of $\vec{j}_1$ and $\vec{j}_2$, and n_α the wobbling of the rotor. Thus, a physical state is labeled by a set of non-negative integers $(n_\alpha, n_\beta, n_\gamma)$, i.e., asymptotic quantum numbers. The lowest band (TSD1) with I has quantum number of (0,0,0), while the second lowest band (TSD2) with $I-1$ has (1,0,0).

Now, we apply this formalism to odd-odd nucleus ^{164}Lu. We compare theoretical energy levels of $E^* - aI(I+1)$ with experimental ones[4,5] in Fig.1. We adopt $i_{13/2}$ orbital both for proton and neutron levels for the positive TSD3 (yrast) band and TSD2 (yrare) band. We choose the parameter set as β_2 =0.38, $\gamma = 17°$, $\mathcal{J}_0 = 77$MeV^{-1} and $c_1 = -4$, $c_2 = 29$ in Eq. (1). The strength of the single-particle potential is $V_{j_1} = V_{j_2} = 2.3$MeV, and the attenuation factor in the Coriolis interactions in $H_{\rm rot}$ is 0.5 common both for proton and neutron terms. The theoretical energy value is adjusted to the experimental one at $I = 17^+$. We see that the overall trend of experimental level schemes is well simulated by this parametrization except for the detailed behavior of those in low spin region ($13 \leq I \leq 20$) and in high spin region ($I \geq 48$). These features are related to the microscopic evolution of nu-

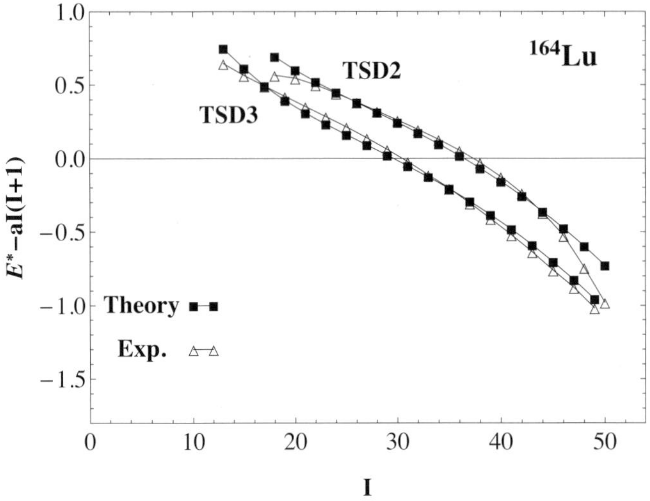

Fig. 1. The comparison between the experimental and the theoretical energy levels, $E^* - aI(I+1)$ as functions of angular momentum I for ^{164}Lu.

clear structure. The same parameter set predicts gradually decreasing out-of-band $E2$ transition rate, $B(E2)_{\rm out} = 0.26 \sim 0.20$ in unit of $(eb)^2$, and gradually increasing in-band transition rate, $B(E2)_{\rm in} = 1.16 \sim 1.19$, in the region of $24 \leq I \leq 32$. The out-of-band $M1$ transition rate is also decreasing with I and much smaller because of the cancellation of spin g-factor, i.e., $B(M1)_{\rm out} = 0.0054 \sim 0.0037$ in unit of $(\mu_{\rm N})^2$ for $24 \leq I \leq 32$. Compared with $B(E2)_{\rm out} = 0.5 \sim 0.4$, $B(E2)_{\rm in} \sim 2.4$ and $B(M1)_{\rm out} = 0.024 \sim 0.014$ in the region of $47/2 \leq I \leq 63/2$ for odd-A ^{163}Lu nucleus,[2] large reduction of $B(M1)_{\rm out}$ is remarkable in the present odd-odd case.

References

1. K. Tanabe, and K. Sugawara-Tanabe, *Phys. Rev. C* **73**, 034305 (2006).
2. K. Tanabe, and K. Sugawara-Tanabe, *Phys. Rev. C* **77**, 064318 (2008).
3. K. Sugawara-Tanabe, and K. Tanabe, *Phys. Rev. C* **82**, 051303(R) (2010).
4. S. Törmänen *et al*, *Phys. Lett. B* **454**, 8 (1999).
5. P. Bringel *et al*, *Phys. Rev. C* **75**, 044306 (2007).

Structure beyond the neutron drip-line: ^{9}He

T. Al Kalanee[1,2], J. Gibelin[1,2], P. Roussel-Chomaz[2], D. Beaumel[3], Y. Blumenfeld[3], B. Fernandez-Dominguez[2], C. Force[2], L. Gaudefroy[2], A. Gillibert[4], J. Guillot[3], H. Iwasaki[3], N. Keeley[6], S. Krupko[5], V. Lapoux[4], W. Mittig[2], X. Mougeot[4], L. Nalpas[4], N. A. Orr[1], E. Pollacco[4], K. Rusek[6], T. Roger[2], H. Savajols[2], N. de Séréville[3], S. Sidorchuk[5], D. Suzuki[3] and I. Strojek[6]

[1]*LPC-ENSICAEN, IN2P3-CNRS et Université de Caen, F14050 Caen Cedex, France*
[2]*GANIL, CEA/DSM-CNRS/IN2P3, BP55027, F14076 Caen Cedex 5, France*
[3]*IPN Orsay, IN2P3-CNRS, Université Paris Sud, F91406 Orsay, France*
[4]*CEA-Saclay, DSM/IRFU SPhN, F91191 Gif-sur-Yvette Cedex, France*
[5]*Flerov Laboratory of Nuclear Reactions, JINR, Dubna, RU141980, Russia*
[6]*Andrzej Sołtan Institute for Nuclear Studies, ul. Hoza 69, 00-681 Warsaw, Poland*

In order to solve the nature of ^{9}He ground state, additional information on this unbound nucleus with extreme N/Z ratio was needed. The present study was performed *via* the (d,p) reaction, a standard tool for determination of neutron single-particle distribution.

Keywords: nuclear structure, direct reaction, virtual states, ^{9}He

1. Introduction

Recent attempts to study the neutron unbound ^{9}He isotope have provided a number of conflicting observations concerning its level structure.[1–4] One of the principal issues is the character of the ground-state and whether the parity inversion in the less exotic $N = 7$ isotones, as predicted by Talmi and Unna,[5] occurs in ^{9}He. In order to shed light on the level structure of ^{9}He, we have performed the ^{8}He(d,p)^{9}He reaction in inverse kinematics using the newly commissioned MUST2 Si-CsI telescope array.

2. Experimental Setup

The experiment was performed at the GANIL facility. The secondary ^{8}He beam at 15.4 A.MeV was produced by the SPIRAL1[6] facility *via* the reaction of a $\sim$ 75 A.MeV ^{13}C beam into a thick carbon target. The beam was delivered to solid deuterated polypropylen targets (CD_2) 320 μg/cm^2

or 546 μg/cm^2 thick, located at the scattering chamber of the SPEG spectrometer.

The beam spot on the target and incident angles were monitored by two sets of multi-wire low pressure chambers, CATS,[7] placed upstream of the target. The energies and angles of the recoiling protons were measured by an array of four MUST2 telescopes[8] located 15 cm upstream of the target. Each telescope, with an active area of 10×10 cm^2, consisted of a 0.3-mm thick double-sided Si strip detector (DSSD) and a 4-cm thick 16-fold CsI calorimeter, which provided energy-loss (ΔE) and residual-energy (E) measurements, respectively. The DSSD was divided into 128 strips in both x and y directions, providing position information on the array. The emission angle of the recoiling particles was thus obtained by combining the information on the beam.

The setup covered laboratory (center-of-mass) scattering angles lab (C.M.) of 120°–180°(0°-21.2°). The acceptance of the array for the present reaction was estimated by Monte-Carlo simulation, which took into account both detector geometry and beam profile. The acceptance has a maximum value of ~80% at lab angles between 130°–160°(C.M. ~ 6°–16°), while it gradually decreases toward smaller or larger angles. Particle identification was achieved using the $E-$TOF method.

The beam particles were detected by a 20×20 mm^2 plastic scintillator located 11 cm downstream of the target and covering the most forward angles up to 5.2°. Higher angles up to 6.5° were covered by a fifth MUST2 telescope located at 19 mm behind the plastic. Finally a sixth MUST2 module was placed at 15 cm, down to the target, away from the beam by an angle of 65° to measure the elastic scattering.

For more details on the experimental setup the reader is referred to Ref. 9–11.

3. Results

3.1. *Test experiment*

The ^{16}O(d,p)^{17}O test reaction at 15.5 A.MeV in reverse kinematics, was performed to validate our understanding of the setup and the analysis procedures. Several single particle states of the ^{17}O were observed: the ground state (G.S.) and 870, 5084 and 5697 keV excited states with an overall deviation from tabulated energies smaller than 5 keV. Experimental resolution was measured between 250 keV (FWHM) for the G.S. to 430 keV to the highest excited states.

Using CH89[12] optical potential parameterization, standard Distorted Wave Born Approximation (DWBA) calculations were carried out using the FRESCO[13] code. For all the states of interest they well reproduce the experimental angular distributions. From them spectroscopic factors were deduced within 30% of the previously published ones, see Table 1.

Table 1: Comparison between our test results and tabulated energy E and spectroscopic factor S for single particle states in ^{17}O.

	Tabuled values and Ref 14,15		This work	
J^π	E(keV)	S	E(keV)	S
$5/2^+$	0	0.84 – 1.07	5±2	0.7
$1/2^+$	870±1	0.91 – 1.14	865±9	1.4
$3/2^+$	5084±9	1.2	5089±1	0.8
$7/2^-$	5697±4	0.15	5692±7	0.13

All these results confirmed the methodological validity of our approach.

3.2. *Helium-9*

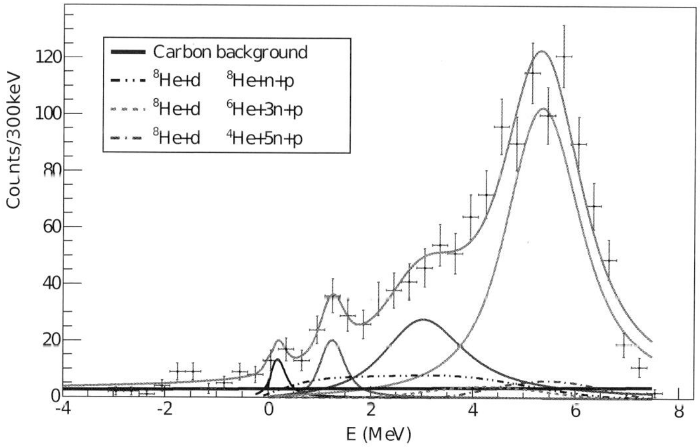

Fig. 1: Missing mass spectrum for the ^{8}He(d,p) reaction.

The analysis method for the reaction on ^{8}He was identical to the ^{16}O one. The missing mass spectrum of ^{9}He presented in Fig. 1 is obtained in coincidence with the identification of an ^{8}He in the plastic beam dump.

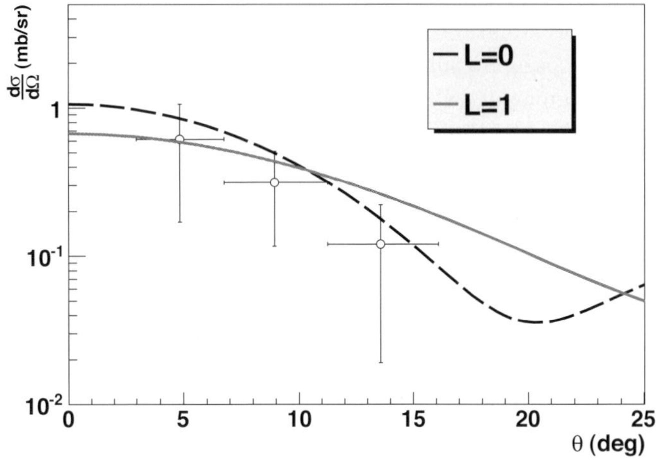

Fig. 2: Angular distribution for the ground state of ^{9}He.

This spectrum exhibits a structure around the ^{8}He neutron threshold, and two others around 1.3 MeV and 3 MeV. The large structure observed in the spectrum around 6 MeV (magenta line) is produced by the acceptance limitation of the experimental response of the detector. We analyzed the result assuming a G.S. near the neutron emission threshold and two excited states. The experimental resolutions obtained from the ^{17}O spectrum were included in the fit. Phase spaces of 3, 5 and 7-body were taken into account as well as the backgrounds due to the reactions of the beam with the carbon of the target and induced reactions in the plastic beam stopper.

Results from previous neutron transfer experiments on ^{8}He[1,2] were used to test the consistency of our obtained resonance parameters. The angular distributions were compared with DWBA calculations, where the ground state near the neutron threshold was considered weakly bound. We present experimental results in Fig. 2 for the ground state compared with theoretical $L = 0$ and $L = 1$ distributions. Results for the other states are also available.

4. Discussion

We focus in this article only on the G.S. of ^{9}He: considering the limited statistics and angular coverage, it is difficult to clearly assign a multipolarity from the angular distribution presented in Fig. 2. We hence performed a multipole decomposition analysis, taking into account the background angular distribution coming from events presenting excitation energies lower than the G.S. ones. This complementary analysis showed that any contribution from $L = 1$ should be at least an order of magnitude lower that the $L = 0$ one, thus confirming the $s_{1/2}$ nature of the ^{9}He ground state. Further details on the multipole decomposition analysis as well as the significance of the spectroscopic factors obtained using more complete CRC calculations will be given in a forthcoming publication.[16]

5. Conclusion

We performed a one-neutron transfer reaction on the neutron rich ^{8}He in order to clarify the structure of the unbound ^{9}He. Our results show a ground state located 0.2 ± 0.04 MeV above the neutron threshold, with a width of 0.2 ± 0.49 MeV. From the results obtained by two different analysis performed on the corresponding angular distribution, we conclude that our results seem to be more in favor of a $s_{1/2}$ state. Two other states, not discussed here, were found at around 1.3 and 3 MeV.

6. Acknowledgments

We wish to thank the GANIL technical staff. We wish to thank R. Raabe for fruitful discussions.

References

1. M. S. Golovkov *et al.*, *Phy. Rev. C* **76**, p. 021605 (2007).
2. S. Fortier *et al.*, AIP Conf. Proc. Vol. 912 p. 3, (2007).
3. H. T. Johansson *et al.*, *Nucl. Phys. A* **842**, p. 15 (2010).
4. L. Chen *et al.*, *Phys. Lett. B* **505**, p. 21 (2001).
5. I. Talmi and I. Unna, *Phy. Rev. Lett.* **4**, p. 469 (1960).
6. A. C. C. Villari, *Nucl. Phys. A* **693**, p. 465 (2001).
7. S. Ottini *et al.*, *Nucl. Instr. Meth. A* **431**, p. 476 (1999).
8. E. Pollacco *et al.*, *Eur. Phys. J. A* **25**, p. 287 (2005).
9. T. Al Kalanee, PhD thesis, Univ. Caen, (Univ. Caen, 2010). and ref. therein, `http://tel.archives-ouvertes.fr/tel-00557105/`.
10. D. Suzuki, PhD thesis, Univ. Tokyo, (Univ. Tokyo, 2009).
11. D. Suzuki *et al.*, *Phys. Rev. Lett.* **103**, p. 152503 (2009).
12. R. L. Varner *et al.*, *Phys. Rep.* **201**, p. 57 (1991).
13. I. J. Thompson, *Comp. Phys. Rep.* **7**, p. 167 (1988).
14. M. D. Cooper *et al.*, *Nucl. Phys. A* **218**, p. 249 (1974).
15. S. E. Darden *et al.*, *Nucl. Phys. A* **208**, p. 77 (1973).
16. T. Al Kalanee *et al.* To be submitted to Phys. Rev. C.

TENSOR CORRELATION IN LIGHT NUCLEI STUDIED WITH TENSOR OPTIMIZED SHELL MODEL

Takayuki Myo

General Education, Faculty of Engineering, Osaka Institute of Technology, Osaka, Osaka 535-8585, Japan
E-mail: myo@ge.oit.ac.jp

Atsushi Umeya

Human Science and Common Education, Faculty of Engineering, Nippon Institute of Technology, Saitama 345-8501, Japan

Hiroshi Toki

Research Center for Nuclear Physics (RCNP), Osaka University, Ibaraki, Osaka 567-0047, Japan

Kiyomi Ikeda

RIKEN Nishina Center, Wako, Saitama 351-0198, Japan

We propose a new theoretical approach to describe nucleus using bare nuclear interaction, in which the tensor and short-range correlations are described with the tensor optimized shell model (TOSM) and the unitary correlation operator method (UCOM), respectively. We show the obtained results of He isotopes using TOSM+UCOM, such as the importance of the pn-pair correlated by the tensor force, and the structure differences in the LS partner states of ^{5}He.

Keywords: bare interaction, tensor correlation, short-range correlation.

1. Introduction

The nucleon-nucleon interaction has distinctive features. There exist strong tensor interaction at intermediate distance caused by pion exchange and strong repulsion at a short distance caused by quark dynamics. It is important to investigate the nuclear structure in relation with the above characteristics of the nucleon-nucleon interaction. There are two important developments for this purpose. One is to find out that the strong tensor force is of intermediate range and we are able to express the tensor correlation in a reasonable shell model space.[1,2] We name this method as Tensor Optimized Shell Model (TOSM). In TOSM, the wave function is constructed in terms of the shell model basis states with full optimization of the two particle-two hole ($2p2h$) states, in particular, the spatial shrinkage of the particle states induced by the tensor force. The other is the Unitary Correlation Operator Method (UCOM) to treat the short-range correlation.[3] We shall combine two methods, TOSM and UCOM, to describe nuclei using bare interaction and apply this new method to light nuclei and see how this method works. We have also investigated the characteristics of the tensor correlations in He and Li isotopes, such as the p-wave splitting energy in ^{5}He,[4] and the neutron halo formation in ^{11}Li.[1]

2. Tensor Optimized Shell Model (TOSM)

We begin with a many-body Hamiltonian having the bare nucleon-nucleon interaction, AV8′.[5] We explain the TOSM wave function Ψ as

$$|\Psi\rangle = \sum_{k_0} A_{k_0}|0p0h\rangle_{k_0} + \sum_{k_1} A_{k_1}|1p1h\rangle_{k_1} + \sum_{k_2} A_{k_2}|2p2h\rangle_{k_2}. \qquad (1)$$

Here, $|0p0h\rangle$, $|1p1h\rangle$ and $|2p2h\rangle$ are the $0p0h$, $1p1h$ and $2p2h$ basis states, respectively, with various radial components for particle states distinguished by the labels k_0, k_1 and k_2. The variational coefficients are given as $\{A_{k0}, A_{k1}, A_{k2}\}$. The hole states are described by harmonic oscillator basis states. We use the Gaussian expansion method to describe the various radial basis states in the particle states.[2] We construct the ortho-normalized single-particle wave function using a linear combination of Gaussian bases $\{\phi_\alpha\}$ for the orbit α with the length parameter $b_{\alpha,\nu}$ as

$$\phi_\alpha(\vec{r}, b_{\alpha,\nu}) = N_l(b_{\alpha,\nu})\; r^l\; e^{-(r/b_{\alpha,\nu})^2/2}\; [Y_l(\hat{r}), \chi^\sigma_{1/2}]_j\; \chi_{t_z}, \qquad (2)$$

where ν is an index that distinguishes the bases with different lengths $b_{\alpha,\nu}$. The quantum numbers l and j are the orbital and total angular momenta of the single-particle states, respectively, and t_z is the isospin component. The normalization factor is denoted as $N_l(b_{\alpha,\nu})$. We superpose a sufficient number of Gaussian bases with various length parameters, which fully optimizes the radial components of the particle states in each configuration independently. We eliminate the center-of-mass excitations using the projection technique of the lowest HO state for the center-of-mass motion.

We use UCOM[3] to treat the short-range correlation in the TOSM basis states. In UCOM, the shift operator is introduced for every nucleon pair in nuclei. This operator reduces the short-range amplitude of the relative pair wave functions. The amount of the shifts is determined variationally.[2]

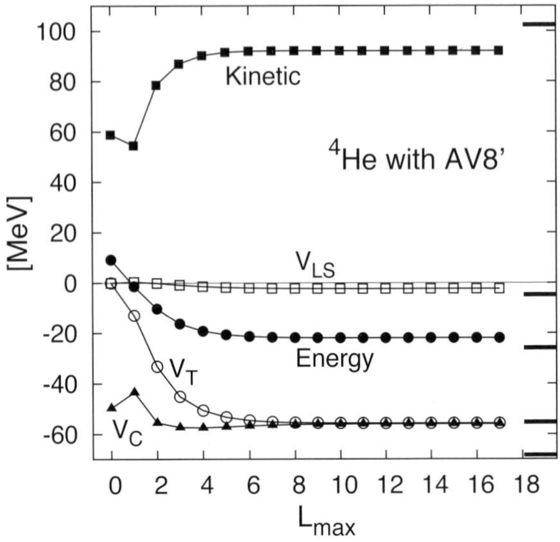

Fig. 1. Components of the central (V_C), tensor (V_T) and spin-orbit (V_{LS}) forces, the kinetic energy (Kinetic) and total energy (Energy) for ^{4}He as functions of $L_{\max}$. The benchmark results[6] are also shown by the short lines on the right-hand side of the figure.

3. ^{4}He and ^{5}He in TOSM + UCOM

We show the results for ^{4}He using TOSM with UCOM in Fig. 1, where $L_{\max}$ is the maximum orbital angular momentum of the particle states. We get a good convergence of the energy with respect to $L_{\max}$ and compare the obtained results with rigorous calculations.[6] We show the components of the energy for ^{4}He in Fig. 1. A very interesting feature is the kinetic energy, which goes up to a large value as the tensor contribution becomes large. We also see that V_C satisfies the rigorous value, -55 MeV. On the other hand, the tensor force matrix element V_T converges and the difference from the rigorous value is about 13 MeV. The kinetic energy difference is also approximately 10 MeV. As the net value, the total energy is -23 MeV which is close to the rigorous value -26 MeV. One of the possibilities to explain the lack of the energy in this approach is the treatment of the short-range part of the tensor correlations. Although the dominant part of the tensor force is of intermediate and long range, there remains small strength in the short-range part of the tensor force, which can couple with the short-range correlations. From Table 1, in ^{4}He, the $0p0h$ configuration exhausts the 83 % of the wave function. It is found that several $2p2h$ states such as

Table 1. Mixing probabilities in the ^{4}He ground state in %, where the two subscripts are the spin-isospin quantum numbers.

$(0s)^4_{00}$	83.01
$(0s)^{-2}_{10}(0p_{1/2})^2_{10}$	2.45
$(0s)^{-2}_{10}[(1s_{1/2})(0d_{3/2})]_{10}$	2.31
$(0s)^{-2}_{10}[(0p_{3/2})(0f_{5/2})]_{10}$	1.90
$(0s)^{-2}_{10}[(0p_{1/2})(0p_{3/2})]_{10}$	1.45

Table 2. Hamiltonian component in ^{5}He measured from those of the ^{4}He ground state. Units in MeV where $\hbar\omega = 18.43$ MeV.

^{5}He(J^π)	Total Energy	Kinetic	Central	Tensor	LS
$3/2^-$	6.97	24.14	−8.99	−5.60	−2.58
$1/2^-$	10.05	17.53	−6.96	−1.11	1.04

$(0s)^{-2}_{10}(0p_{1/2})^2_{10}$ are largely mixed and those configurations are essential for the tensor correlation in ^{4}He.[2]

We have applied TOSM+UCOM to ^{5}He and investigate the effect of tensor force on the splitting energy of ^{5}He. In Table 2, we compare the Hamiltonian components in two states of ^{5}He measured from those of ^{4}He. In the calculation, we exclude the continuum effect and discuss the internal structure of ^{5}He. The obtained LS splitting energy is about 3 MeV. It is found that the energy of the $3/2^-$ state has a larger tensor contribution than that of the $1/2^-$ state. This is related to the large mixing of the $p_{1/2}$ component in ^{4}He shown in Table 1. When the last neutron in ^{5}He occupies the $p_{3/2}$ orbit, this neutron does not disturb the ^{4}He structure. In case of the $p_{1/2}$ occupation in the $1/2^-$ state, this neutron produces the Pauli-blocking effect and reduces the total energy of ^{5}He. The large tensor contribution in ^{5}He($3/2^-$) brings the enhancement of the kinetic energy beyond $\hbar\omega$ owing to the high momentum nature of the tensor force. The same results of the role of tensor force as the above one can be seen in ^{6,7,8}He.

References

1. T. Myo, K. Katō, H. Toki and K. Ikeda, Phys. Rev. **C76**, 024305 (2007).
2. T. Myo, H. Toki and K. Ikeda, Prog. Theor. Phys. **121**, 511 (2009).
3. H. Feldmeier, T. Neff, R. Roth and J. Schnack, Nucl. Phys. **A632**, 61 (1998).
4. T. Myo, K. Katō, K. Ikeda, Prog. Theor. Phys. **113**, 763 (2005).
5. S.C. Pieper and R. B. Wiringa, Annu. Rev. Nucl. Part. Sci. **51**, 53 (2001).
6. H. Kamada et al., Phys. Rev. **C64**, 044001 (2001).

Symmetry Energy, Pairing correlations in Nuclear Matter and Giant Resonances in Tin Isotopes

Myung-Ki Cheoun, Eunja Ha, and C. Y. Ryu

Department of Physics, Soongsil University, Seoul 156-743, Korea
**E-mail: cheoun@ssu.ac.kr*

We calculated isoscalar giant monopole and isovector giant dipole resonances (GDRs) in tin isotopes using the quasi-particle random phase approximation (QRPA). Centroid energies for the giant resonances are derived from relevant strength distributions and compared to the available data. Roles of the pairing energies in the giant resonances are also discussed in detail.

Keywords: Symmetry Energy, Giant Monopole and Pygmy Dipole Resonances

1. Introduction

The symmetry energy of the nuclear systems is one of the hot topics in the nuclear physics as well as in the nuclear astro physics because it plays an important role in constraining the equation of state of nuclear matters.[1] Since the symmetry energy is given as the difference of pure neutron matter (PNM) and symmetric nuclear matter (SNM),[2] the exact information is of physical importance to understand the structure of neutron stars as well as neutron rich nuclei.

The symmetry energy has actually two independent variables, density dependence and asymmetry dependence given as $(N-Z)/A$ or $(\rho-\rho_0)/3\rho_0$ with the density ρ_0 at a saturation density.[2] While the asymmetry dependence could be realized in nuclear data, the density dependence is not fully understood because the density dependence of nucleon-nucleon interaction beyond the saturation density is still unknown.

The asymmetry dependence appears also at the pairing energy which stems from the pairing correlations remained still after the conversion of the nucleon-nucleon interactions into the mean field. For example, the mass excess of N = Z nuclei should not have any N or Z dependence, but it shows a zigzag dependence on total mass number A. Therefore the pairing energy should be also considered for the proper account of the symmetry energy.

In this talk, we focus on the giant resonances, such as GDRs and their pygmy types, and giant monopole resonance (GMR) because the collective motions of the nuclear system could give a unique chance of deducing information of the symmetry energy in finite nuclei.

In this work, we perform evaluations of such resonances using the QRPA formulas, which include neutron-neutron and proton-proton pairing as well as neutron-proton pairing correlations.

2. The Giant Resonances in the QRPA

Here we briefly explain our QRPA formalism.[3] Matrix elements of any transition operator $\hat{O}_\lambda$ between a ground state and an excited state $|\omega; JM>$ are expressed as follows

$$< \omega; JM||\hat{O}_\lambda||QRPA >= [\lambda]^{-1} \sum_{ab} < a||\hat{O}_\lambda||b >< QRPA||[c_a^+ \tilde{c}_b]_\lambda||\omega; JM > , \quad (1)$$

where the first factor $< a||\hat{O}_\lambda||b >$ can be calculated independently of nuclear models for a given single particle basis. Excited states, $|m; J^\pi M\rangle$, in a compound nucleus are generated by operating the following one phonon operator to the initial nucleus

$$Q_{JM}^{+,m} = \sum_{kl\mu' \nu'} [X^m_{(k\mu' l\nu' J)} C^+(k\mu' l\nu' JM) - Y^m_{(k\mu' l\nu' J)} \tilde{C}(k\mu' l\nu' JM)] \, , \quad (2)$$

with a pair creation operator C^+. By using the phonon operator $Q_{JM}^{+,m}$, we obtain the following expressions for charge conserving electro-magnetic transitions

$$< \omega; JM||\hat{O}_\lambda|| \, QRPA >= \sum_{a\alpha' b\beta'} [\mathcal{N}_{a\alpha' b\beta'} < a\alpha' ||\hat{O}_\lambda||b\beta' > \quad (3)$$

$$[u_{pa\alpha'} v_{pb\beta'} X_{a\alpha' b\beta'} + v_{pa\alpha'} u_{pb\beta'} Y_{a\alpha' b\beta'}] - (-)^{j_a + j_b + J} \mathcal{N}_{b\beta' a\alpha'}$$

$$< b\beta' ||\hat{O}_\lambda||a\alpha' > [u_{pb\beta'} v_{pa\alpha'} X_{a\alpha' b\beta'} + v_{pb\beta'} u_{pa\alpha'} Y_{a\alpha' b\beta'}]] + (p \rightarrow n) \, ,$$

where $\mathcal{N}_{a\alpha' b\beta'}(J)$ is the nomalization factor. In this report, we exploit the following operators

$$\hat{O}(ISGMR : 0^+) = \frac{Z}{A} \sum_{i=1}^{A} r_i^2, \; \hat{O}(ISGDR : 1^-)^{T=1}_{1\mu} = \frac{Z}{A} \sum_{i=1}^{A} r_i^3 Y_{1\mu} \, . \quad (4)$$

3. Results of Giant Monopole and Dipole Resonances

For GMRs we remove the spurious states at low energy states owing to the particle number non-conservation. If we define various magnetic moments

$m_k = \int E^k S(E) dE$ with the strength function $S(E)$ for each operator, the sum rule for the ISGMR operator, with the help of the Thouless theorem $m_1 = (2A\hbar^2/m) < r^2 >$ with nucleon mass m, is derived as follows[4]

$$\sum_k E_k | < k|\hat{O}(ISGMR: 0^+)|0 > |^2 = 2\frac{\hbar^2}{m}\frac{Z^2}{A} < r^2 > \ (MeV fm^4)\ . \quad (5)$$

This sum rule could be a useful test for the nuclear model used. The centroid energy of the ISGMR is defined as the ratio of the energy weighted sum rule (EWSR) to the non-energy weighted sum rule (NEWSR), $E_{GMR} = m_1/m_0$. Our results for the GMR in tin isotopes are tabulated in Table 1. For

Table 1. Various moments of the ratios for the GMR defined in text, $E_1 = \sqrt{m_1/m_{-1}}$, E_{GMR} and $E_3 = \sqrt{m_3/m_1}$.

Sn	112	114	116	118	120	122	124
EWSR	112	114	116	118	120	122	124
E_1	112	114	116	118	120	122	124
E_{GMR}	17.75	14.47	13.25	15.85	17.45	16.22	15.47
E_3	112	114	116	118	120	122	124

GDRs, spurious states arising from the center of mass motion are removed by the effective charges. The energy weighted sum rule (EWSR) satisfied[5] we exploited, known as the TRK sum rule, is given as

$$\sum_k E_k | < k|\hat{O}(IVGDR: 1^-)^{T=1}_{1\mu}|0 > |^2 = \frac{\hbar^2}{2m}\frac{9}{4\pi}\frac{NZ}{A}e^2(MeV^{-1})\ . \quad (6)$$

The TRK sum rule is related to the total photoreaction cross section. Detailed discussions about the relation are referred from Ref.[5] Our results are shown in Fig. 1. The numerical ratio of PDR and GDR are shown Table 2.

Table 2. Various moments of the IVGDR through the energy weighted sum rule (EWSR) with the strength functions of the GDR and PDR

A	PDR			GDR			PDR/GDR	
	m_0	m_1	E_c	m_0	m_1	E_c	m_0 (%)	m_1 (%)
112	0.42	3.05		4.58	93.4		3.26	9.28
114	0.52	3.62		6.16	122.4		2.95	8.50
116	0.98	8.07		4.95	116.3		6.94	19.9
118	1.61	13.9		12.3	213.6		6.49	13.1
120	1.15	9.86		12.66	226.3		4.35	9.09
122	2.92	26.4		13.61	213.8		12.4	21.4
124	3.29	31.9		14.99	230.8		13.8	23.3

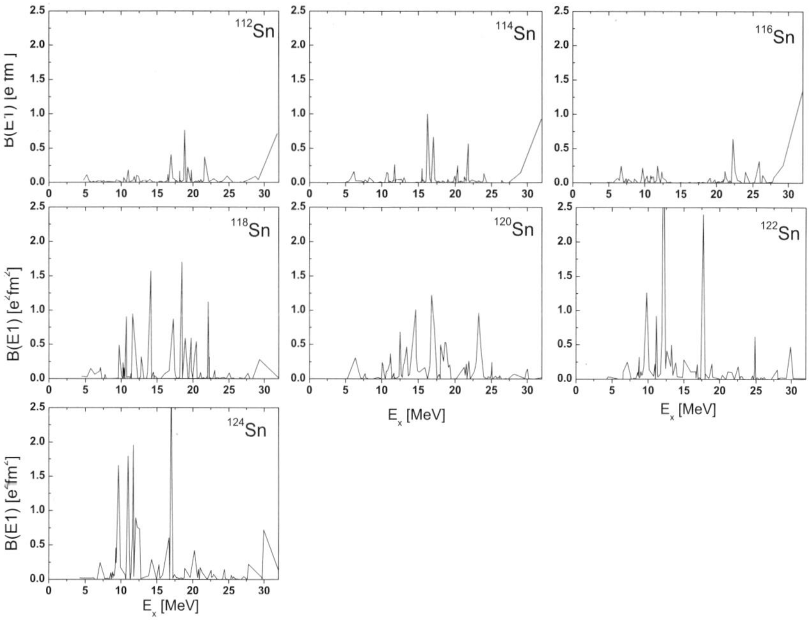

Fig. 1. The E1 strength distributions in tin isotopes. PDR and GDR are verified around 5 MeV and other energy regions, respectively. All pairing interactions are included to reproduce the empirical mass excess.

4. Summary and Conclusion

We estimated giant monopole and dipole resonances in tin isotopes using a QRPA. Our results show explicitly the isospin dependence for GMR and GDR. The appearance of pygmy type resonances in the GDR is also shown explicitly. More synthetic calculations for studying roles of pairing correlations in the symmetry energy for finite nuclei are in progress.

References

1. B. -A. Li, L. -W. Chen and C. M. Ko, Phys. Rep. **464**, 113 (2008).
2. J. Piekarewicz and M. Centelles, Phys. Rev. C **79**, 054311 (2009).
3. Myung-Ki Cheoun, Eunja Ha, K. S. Kim and T. Kajino, J. of Phys. **G 37**, 055101, (2010).
4. M. Centelles, S. K. Partra, X. Roca-Maza, P. D. Stevenson and X Vinaes, J. Phys. G. **37**, 075107 (2010).
5. H. Sagawa and Toshio Suzuki, Phys. Rev. C **59**, 3116 (1999).

SYSTEMATIC STUDY OF $E1$ MODE USING CANONICAL-BASIS TDHFB

S. EBATA

Center for Nuclear Study, University of Tokyo,
2-1 Hirosawa, Wako-shi, Saitama 351-0198, Japan
E-mail: ebata@cns.s.u-tokyo.ac.jp

T. NAKATSUKASA and T. INAKURA

RIKEN Nishina Center,
2-1 Hirosawa, Wako-shi, Saitama 351-0198, Japan

K. YOSHIDA

Department of Physics, Graduate School of Science and Technology,
Niigata University, Niigata 950-2181, Japan.

Y. HASHIMOTO and K. YABANA

Graduate School of Pure and Applied Sciences, University of Tsukuba,
Tsukuba, Ibaraki 305-8571, Japan

We systematically investigate the $E1$ strength distributions of even-even nuclei ($Z = 8 - 20$ isotopes) using the canonical-basis time-dependent Hartree-Fock-Bogoliubov theory that can describe the dynamics of superfluid nuclei. Especially, we focus on the low-lying $E1$ strength that is often called pygmy dipole resonance and investigate the neutron-number dependence of the strength in neutron-rich nuclei. The evolution of the low-lying $E1$ strength is determined by several factors including the neutron-shell effects, the deformation and the pairing.

Keywords: Neutron-rich nuclei, Pygmy dipole resonance, Pairing correlation

1. Introduction

The region of unstable nuclei reachable by recent experimental technique is rapidly expanding. In order to analyze and predict the properties of unstable nuclei, theoretical models capable of describing properties of their ground states and excited states are required. For this purpose, we have developed the canonical-basis time-dependent Hartree-Fock-Bogoliubov theory (Cb-TDHFB) and applied it to several unstable nuclei which are deformed in a superfluid phase[1].

In this work, we focus on the low-lying $E1$ mode that is often called pygmy dipole resonance (PDR). The PDR is known as one of the most characteristic excited modes in neutron-rich nuclei. The PDR was predicted as a soft-dipole mode for neutron-halo nuclei such as ^{11}Li [2,3] and is expected to play an important role in the nucleosynthesis[4]. We systematically investigate the low-lying $E1$ strength of neutron-rich nuclei in order to understand the basic properties of the PDR.

2. Cb-TDHFB equations and numerical details

The Cb-TDHFB can be derived from the TDHFB equations on the assumption that the pair potential can be approximated to be diagonal in the canonical-basis. In numerical calculations, we adopt the Skyrme interaction (SkM*) and a constant G pairing functional, $E_{\rm pair} = -G|\sum_{l>0}\kappa_l(t)|^2$, then they are written as

$$i\hbar\dot{\phi}_l(\boldsymbol{r},t) = \big(h[\rho(t)] - \varepsilon_l(t)\big)\phi_l(\boldsymbol{r},t), \quad i\hbar\dot{\phi}_{\bar{l}}(\boldsymbol{r},t) = \big(h[\rho(t)] - \varepsilon_{\bar{l}}(t)\big)\phi_{\bar{l}}(\boldsymbol{r},t), \quad (1)$$

$$i\hbar\dot{\rho}_l(t) = \Delta^*(t)\kappa_l(t) - \Delta(t)\kappa_l^*(t), \quad (2)$$

$$i\hbar\dot{\kappa}_l(t) = (\varepsilon_l(t) + \varepsilon_{\bar{l}}(t))\kappa_l(t) + \Delta(t)(2\rho_l(t) - 1), \quad (3)$$

where $\phi_l(\boldsymbol{r},t)$ and $\phi_{\bar{l}}(\boldsymbol{r},t)$ respect a pair of canonical states which are not necessarily relate to each other by the time reversal, $|\phi_{\bar{l}}\rangle \neq T|\phi_l\rangle$. $\rho_l(t) = |v_l(t)|^2$ and $\kappa_l(t) = u_l(t)v_l(t)$ are occupation- and pair-probabilities, given in terms of the time-dependent BCS factor $(u(t), v(t))$. The gap energy is defined as $\Delta(t) \equiv G\sum_{k>0}\kappa_k(t)$, and $\varepsilon_l(t) \equiv \langle\phi_l(t)|h|\phi_l(t)\rangle$. These equations conserve the orthonormality of the single-particle wave functions, the average particle number, and the average total energy.

We solve the Cb-TDHFB equations in real time and real space, then calculate the linear response of the nucleus. We calculate the time evolution of the expectation value of a one-body operator $\hat{F}$ after adding an instantaneous external field, $V_{\rm ext}(t) = -k\hat{F}\delta(t)$ with $k \ll 1$, to the ground state and obtain the strength function $S(\hat{F};E)$ by Fourier transformation. The $E1$ operator is adopted for the operator $\hat{F}$. In the present calculation, we use the three-dimensional Cartesian coordinate-space representation for the canonical states, $\phi_l(\boldsymbol{r},\sigma;t) = \langle\boldsymbol{r},\sigma|\phi_l(t)\rangle$ with $\sigma = \pm 1/2$. The coordinate space is discretized in the square mesh of $\Delta x = \Delta y = \Delta z = 0.8$ fm in a sphere with radius of 12 fm. The details of this method can be found in Ref.[1].

3. Results and Discussion

We define the following ratio in order to quantify the low-lying $E1$ strength;

$$\frac{m_1(E_c)}{m_1} \equiv \frac{\int^{E_c} E \times S(E1;E)dE}{\int E \times S(E1;E)dE}, \tag{4}$$

where the energy cutoff E_c =10 MeV is taken in this study.

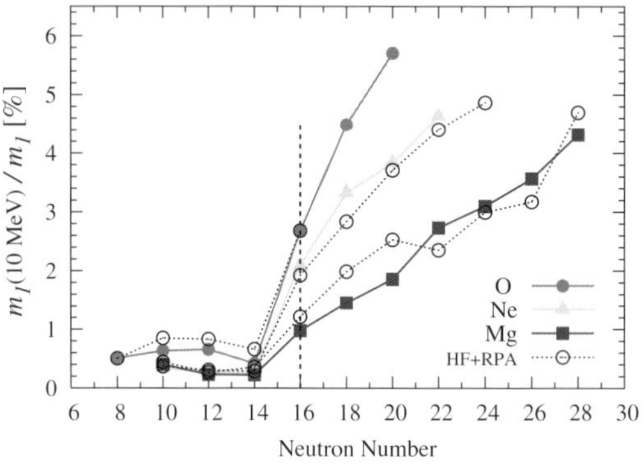

Fig. 1. The $m_1(E_c)/m_1$ defined by Eq.(4) for O (filled circle), Ne (triangle) and Mg (square) isotopes as a function of neutron number. The solid lines are results of Cb-TDHFB and the dotted line with circle symbols are results of HF+RPA without pairing correlation.

Figure 1 shows the results for Oxygen(O), Neon(Ne) and Magnesium(Mg) isotopes. The solid lines with filled symbols indicate the results of Cb-TDHFB calculation and the dotted lines with open circles are those of Hartree-Fock plus random phase approximation (HF+RPA) without the pairing correlation[5]. In both cases, we can see the obvious growth of the PDR function at $N = 16$ on each isotopic chain. This suggests that the occupation of particular single-particle orbits is important for the appearance of the low-lying $E1$ strength. We can see same difference between the two calculations with and without pairing for Ne and Mg isotopes with $N = 16, 18, 20$. This is due to different shape in the ground state. Namely, the calculated ground states in the HF+BCS are spherical, but those in the HF are deformed. The pairing mostly affects the low-lying $E1$ strength through the ground-state shape.

Figure 2 shows the results of Silicon(Si), Sulfur(S), Argon(Ar) and Calcium(Ca) isotopes. The low-lying $E1$ strength increases smoothly around $N = 28$ in the S and Ar isotopes in the Cb-TDHFB calculation, while we see a sudden increase from $N = 28$ to 30 in the HF+RPA calculation. This smooth evolution of the PDR strength is caused by the fractional occupa-

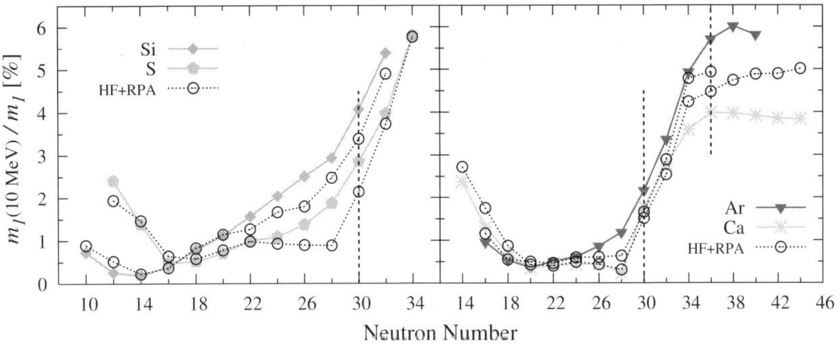

Fig. 2. Same as Fig.1, but for Silicon (diamond) and Sulfur (pentagon), Argon (inverted triangle), and Calcium(star), as a function of neutron number.

tion probabilities of the single-particle orbits, due to the pairing correlation. Beyond the neutron number $N = 36$, the increasing ratio of the PDR function decreases in Ar and Ca isotope.

Figures suggest a shell effect that the property of the occupied single-particle state is important for the appearance of the PDR in neutron-rich nuclei. The neutron number $N = 16$ (30) corresponds to the occupation of $s_{1/2}$ ($p_{3/2}$) orbit, and $N = 36$ corresponds to the occupation of $f_{5/2}$ orbit. The low-angular-momentum orbits such as s and p orbits, enhance the low-lying $E1$ strength, whereas the high-angular-momentum orbits, such as f or higher-l orbits do not.

Currently, we are investigating the low-lying $E1$ modes in the heavier nuclear mass region, to see how they are different from those in light nuclei.

References

1. S. Ebata, T. Nakatsukasa, T. Inakura, K. Yoshida, Y. Hashimoto and K. Yabana, Phys. Rev. C**82**, 034306 (2010).
2. K. Ikeda, Nucl. Phys. A **538** 355 (1992).
3. T. Nakamura *et al.*, Phys. Rev. Lett. **96**, 252502 (2006).
4. S. Goriely, Phys. Lett. B **436** 10 (1998).
5. T. Inakura, T. Nakatsukasa and K. Yabana, AIP Conf. Proc. **1269**, 213 (2010); arXiv:1106.3618.

Studies of light neutron-excess nuclei from bounds to continuum

Makoto Ito
Department of Pure and Applied Physics, Kansai University,
3-3-35 Yamatecho, Suita, Osaka 565-0821, Japan
**E-mail: itomk@kansai-u.ac.jp*

The generalized two-center cluster model (GTCM), which can treat various single particle configurations in general two center systems, is applied to the light neutron-excess system, Be isotopes ($\alpha+\alpha+N+N+$...). We show that the covalent and ionic configurations coexist in a same nucleus.

Keywords: Cluster Structure; Neutron-excess systems; Unbound states.

1. Introduction

In the last two decades, developments of experiments with secondary RI beam have extensively advanced the studies on light neutron-rich nuclei. In particular, much efforts have been devoted to the investigation of molecular structure in Be isotopes. The Be isotopes can be considered as typical examples of two-center superdeformed systems which build on an $\alpha+\alpha$ rotor of ^{8}Be. Theoretically, molecular orbital (MO), such as the π^- and σ^+ orbitals associated with the covalent bonding in atomic molecules, have been successful in understanding the low-lying states of these isotopes.[1]

The MO model can describe many kinds of characteristic properties of these isotopes, but they are mainly limited to the analysis on low-lying bound states, and theoretical studies on the highly excited states above the particle-decay threshold is still open area. In contrast to the situation of theoretical studies, recent experiments on ^{12}Be, for instance, revealed the existence of many resonant states,[2–4] which strongly decay into $^6\mathrm{He}_{g.s.}+{}^6\mathrm{He}_{g.s.}$ and $\alpha+{}^8\mathrm{He}_{g.s.}$. Similar resonances, decaying to He isotopes, have also been observed in other Be isotopes, such as $^{10}\mathrm{Be}=\alpha+{}^6\mathrm{He}$ and $^{14}\mathrm{Be}={}^6\mathrm{He}+{}^8\mathrm{He}$.

In the present study, we investigate the structural changes from the bound states to the continuum states in even Be isotopes. In order to investigate the continuum states above the particle decay threshold, the intrinsic structures and their coupling to the scattering states should be treated in a unified manner. For this purpose, we introduce the generalized two-center cluster model (GTCM).[5,6] In this model, the covalent MO configuration can be smoothly connected to the atomic or ionic one, in which valence neutrons are localized around one of the α cores. Furthermore, it becomes possible to describe both the formation of the covalent MO structures and its decays into a continuum energy state. In this report, we apply GTCM to even Be isotope of $^{8\sim16}$Be, and investigate the structural changes from a bound region to continuum region.

2. Framework

The detailed explanation of GTCM has already been published in Refs. 5,6, and we briefly show the formulation of GTCM in the following. In this model, the total wave function of ^{12}Be is given by the superposition of the basis $\{\Phi^{J^\pi K}_{\mathbf{m}}(S)\}$, where,

$$\Phi^{J^\pi K}_{\mathbf{m}}(S) = \hat{P}^{J^\pi}_K \mathcal{A}\left\{\psi_L(\alpha)\psi_R(\alpha)\prod_{j=1}^{4}\varphi_j(m_j)\right\}_S . \quad (1)$$

The α-cluster $\psi_n(\alpha)$ (n=L, R) is expressed by the $(0\mathrm{s})^4$ configuration of the harmonic oscillator (HO) centered at the left(L)- or right(R)- side with the relative distance S. The single-particle wave function for the four valence neutrons localized around one of the α clusters is given by an atomic orbital (AO) $\varphi(p_k, i, \tau)$, and $0p$-orbitals p_k ($k = x,\ y,\ z$) around i (= L or R) with the spin τ (= $\uparrow$ or $\downarrow$). Here, $\{m_j\}$ are indices of AO (p_k, i, τ) and $\mathbf{m}$ represents a set of AOs for the four neutrons, $\mathbf{m}$=(m_1, m_2, m_3, m_4). The intrinsic basis functions with the full anti-symmetrization $\mathcal{A}$ are projected to the eigenstate of the total spin J, its intrinsic angular projection K, and the total parity π by the projection operator $\hat{P}^{J^\pi}_K$.

The total wave function is finally given by taking the superposition over S, $\mathbf{m}$ and K as

$$\hat{\Psi}^{J^\pi}_\nu = \int dS \sum_{\mathbf{m},K} C^\nu_{\mathbf{m}K}(S)\, \Phi^{J^\pi K}_{\mathbf{m}}(S) . \quad (2)$$

The coefficients for the ν-th eigenstate, $C^\nu_{\mathbf{m}K}(S)$, are determined by solving a coupled channel GCM (Generator Coordinate Method) equation.[7] we include all the possible AO configurations for the four valence neutrons within the axially symmetric (K=0) case.[5,6] As for the nucleon-nucleon interaction, we use the Volkov No.2 and the G3RS for the central and spin-orbit parts, respectively. The parameters in the interactions and the size parameter of HO are the same as those applied in Ref. 6, which successfully reproduce the properties of ^{12}Be.

3. Results

The application of GTCM to $^{12}\text{Be} = \alpha+\alpha+4N$ has already been reported in Ref. 6. In the present report, we mainly show the result of other Be isotopes. The energy spectra of ^{10}Be with $J^{\pi}=0^{+}$ is shown in Fig. 1(a). The total reaction probability, $1 - \Re S_{el}$, which is defined by the scattering matrices of the $\alpha+^{6}\text{He}_{g.s.}$ elastic scattering (S_{el}), is plotted by the dotted curve at the upper part in Fig. 1(a).

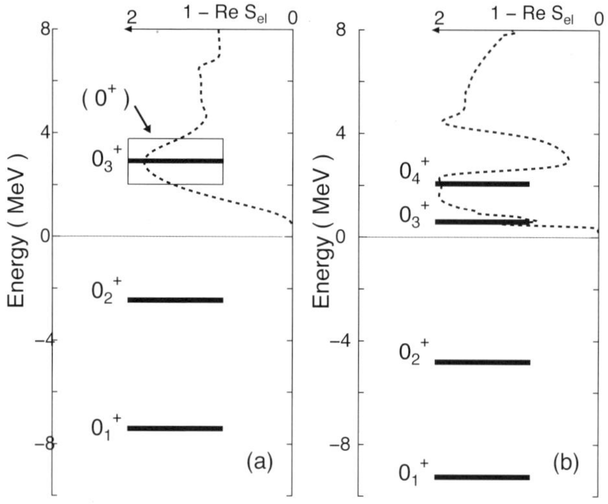

Fig. 1. (a) Energy spectra of ^{10}Be ($J^{\pi}=0^{+}$). The threshold energies of the $\alpha+^{6}\text{He}_{g.s.}$ channel is taken to be the origin. The dotted curve at the right part represents the reaction probability for the central collision of $\alpha+^{6}\text{He}_{g.s.}$. (b) The same figure as (a) but for ^{14}Be. The $^{6}\text{He}_{g.s.}+^{8}\text{He}_{g.s.}$ threshold energy is set to be the origin.

Below the α decay threshold (zero energy point), two bound states appear. The intrinsic structure of the 0_1^+ and 0_2^+ states are nicely described by the MO configuration of $(\pi_{3/2}^{-})^2$ and $(\sigma_{1/2}^{+})^2$, respectively. Since the $\sigma_{1/2}^{+}$ orbit has an enlarged distribution along the α-α axis, the α-α clustering is well developed in the 0_2^+ state. Above the α decay threshold, a prominent peak appears around $E \sim 3$ MeV in the reaction probability. This enhancement of the reaction probability is due to the formation of the resonance with the ionic structure of $\alpha+^{6}\text{He}(2_1^+)$, which is shown by the 0_3^+ state in

Fig. 1(a). At the same energy region, there is a broad continuum strength, shown by 0^+ in parenthesis with a transparent box. This continuum state has a main component of an ionic structure of α+^{6}He$_{g.s.}$. The energy spectra for ^{14}Be is shown in Fig. 1(b). In this figure, we can clearly confirm the similar level structure to ^{10}Be. In the bound region, two 0^+ states appears, while two resonant states are embedded in continuum energy. The ground 0_1^+ state has a spatially compact structure, while the 0_2^+ state has a well developed α-α cluster structure. In the continuum region, the ionic structures are realized; ^{6}He$_{g.s.}$+^{8}He$_{g.s.}$ in 0_3^+ and ^{6}He(2_1^+)+^{8}He$_{g.s.}$ in 0_4^+.

We have also analyzed the level structure of ^{16}Be and found that the ground state in ^{16}Be appears below the ^{8}He$_{g.s.}$+^{8}He$_{g.s.}$ threshold by about 4 MeV. This ground state corresponds to the unbound states above the ^{14}Be$_{g.s.}$+2n threshold by about 3 MeV. The ground state in ^{16}Be, which is predicted by the GTCM calculation, can be considered as an analog state of α+α in ^{8}Be.

4. Summary

In summary, we have performed the systematic calculations of GTCM for even Be isotopes. We have found that, in ^{10,14}Be, the ionic structures are realized as resonances embedded in continuum. In ^{16}Be, an energy level, which can be considered as the ^{8}Be=α+α analog state, is obtained. From the systematic studies, we obtained a wide variety of cluster structures in even Be isotopes. Such various structures can also be expected in other neutron excess systems, such as O isotopes (α+^{12}C+XN) and Ne isotopes (α+^{16}O+XN). Systematic studies are now underway.

References

1. N. Itagaki and S. Okabe, Phys. Rev. C **61**, 044306 (2000), and references therein.
2. A. A. Korsheninnikov et al., Phys. Lett. B **343**, 53 (1995).
3. M. Freer et al., Phys. Rev. C **63**, 034301 (2001).
4. A. Saito et al., Mod. Phys. Lett. A **25**, 21 (2010).
5. M. Ito, K. Kato and K. Ikeda, Phys. Lett. B**588**, 43 (2004); Makoto Ito, Phys. Lett. B**636**, 293 (2006).
6. M. Ito, N. Itagaki, H. Sakurai, and K. Ikeda, Phys. Rev. Lett. **100**, 182502 (2008);M. Ito and N. Itagaki, Phys. Rev. C **78**, 011602(R) (2008); *ibid*, Phys. Rev. Focus Vol. 22, Story 4 (2008).
7. H. Horiuchi et al., Suppl. Prog. Theor. Phys. **62**, 1 (1977).

TRIAXIAL NUCLEAR MOLECULE $^{28}Si - ^{28}Si$ IN RESONANCES

E. UEGAKI

Graduate School of Engineering and Resource Science, Akita University, Tegata, Akita City 010-8502, Japan
E-mail: uegaki@phys.akita-u.ac.jp

Y. ABE

Research Center for Nuclear Physics, Osaka University, 10-1 Mihogaoka, Ibaraki, Osaka 567-0047, Japan
E-mail: abey@rcnp.osaka-u.ac.jp

Resonances observed in the $^{28}Si + ^{28}Si$ collision are studied by a molecular model. At high spins, a stable dinuclear configuration is found to be equator-equator touching one. Energy spectrum of the normal modes around the equilibrium is shown. These modes are expected to be the origin of the large number of resonances observed, hence theoretical analyses for the physical quantities are done. Disalignments between the orbital angular momentum and the fragment spins in the resonance state are clarified. Moreover the analyses of the angular correlations indicate a characteristic feature for each normal-mode excitation. Thus it is possible to identify the mode.

Keywords: Molecular resonances; $^{28}Si + ^{28}Si$ collision; Angular correlations

1. Introduction

High-spin resonances well above the Coulomb barrier in the $^{24}Mg + ^{24}Mg$ and $^{28}Si + ^{28}Si$ systems exhibit many narrow and prominent peaks correlated among the elastic and inelastic channels, which suggest rather long-lived compound systems and offer intriguing subjects in nuclear physics.[1] Angular correlation measurements for $^{28}Si + ^{28}Si$ at $E_{\rm cm} = 55.8$MeV were made with 4π gamma detectors and have provided crucially important information on the resonance, as follows.[2] 1) Angular distributions in the single and mutual 2^+ excitations indicate a dominance of a single orbital angular momentum $L = J = 38$. 2) The angular correlations between the fragments ^{28}Si and the γ-rays emitted from the fragments indicate that the intrinsic spins of the constituent nuclei are in the reaction plane. Thus, dis-alignments between the orbital angular momentum and the fragment spins are expected as a characteristic feature of the resonance state.

2. Dinuclear Normal Modes of Triaxial Molecule ^{28}Si − ^{28}Si

Assuming a constant deformation and axial symmetry of the constituent nuclei for simplicity, we have seven degrees of freedom $(q_i) = (\theta_1, \theta_2, \theta_3, R, \alpha, \beta_1, \beta_2)$, as illustrated in Fig. 1(a), where (R, θ_2, θ_1) is the relative vector between two ^{28}Si nuclei. As internal degrees of freedom, the orientations of the symmetry axes of two constituent ^{28}Si nuclei are described with Euler angles (α_i, β_i) referring to the molecular axes. α_1 and α_2 are combined into $\theta_3 = (\alpha_1 + \alpha_2)/2$ and $\alpha = (\alpha_1 - \alpha_2)/2$. Consistently with the coordinate system, at first, we introduce a rotation-vibration type wave function $D^J_{MK}(\theta_i)\chi_K(R, \alpha, \beta_1, \beta_2)$, as basis one, where χ_K describes internal motions which should be solved by the molecular model.

The authors found the equilibrium configuration of interacting two oblate nuclei to be the *equator-equator* (E-E) one, which is displayed in the lower portion of Fig. 1(b). Couplings among various molecular configurations are taken into account by the method of normal modes around the equilibrium, which gives rise to the molecular modes of excitation. We obtain *butterfly* and *anti-butterfly* modes with $\beta_+ = (\Delta\beta_1 + \Delta\beta_2)/\sqrt{2}$ and $\beta_- = (\Delta\beta_1 - \Delta\beta_2)/\sqrt{2}$, respectively. Associated with the α-degree, $\alpha = (\alpha_1 - \alpha_2)/2$, *twisting* excitations exist. Details are given in Ref. 3.

In Fig. 2(a) molecular normal modes of ^{28}Si $+^{28}$ Si with spin 38 are displayed, classified with K-quantum numbers. The twisting mode is indicated by (t), while the butterfly by (b), respectively. All those intrinsic excitations exist over the range of spin $J = 34 \sim 40$ in which the E-E configuration is stable, and are expected to be the origin of the narrow resonances.

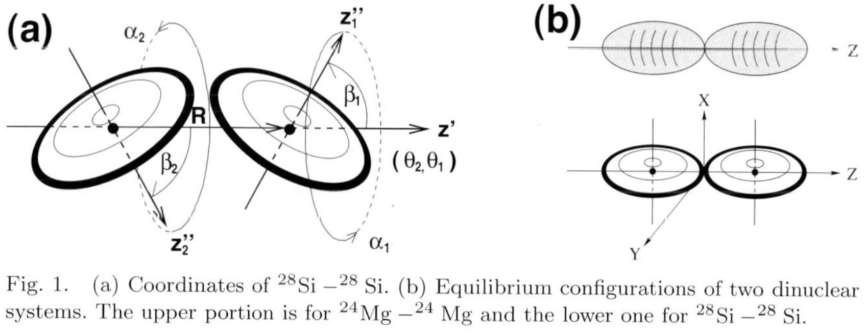

Fig. 1. (a) Coordinates of ^{28}Si $-^{28}$ Si. (b) Equilibrium configurations of two dinuclear systems. The upper portion is for ^{24}Mg $-^{24}$ Mg and the lower one for ^{28}Si $-^{28}$ Si.

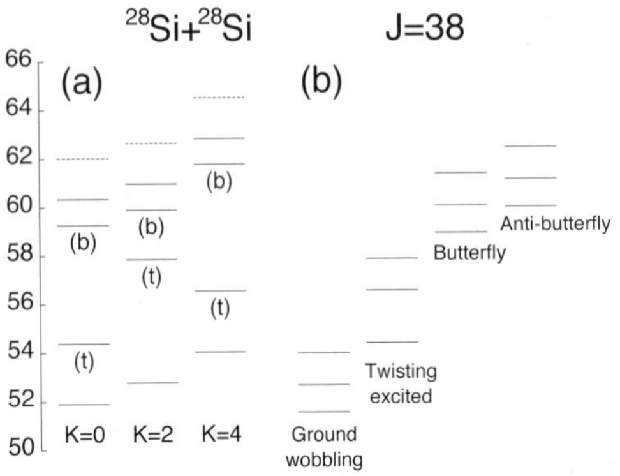

Fig. 2. Energy spectra of the ^{28}Si+^{28}Si system for $J = 38$. (a) Molecular normal modes without K-mixing. (b) After K-mixing, with indications of the modes under the levels.

Since the E-E configuration is *triaxial* as is seen in Fig. 1(b), rotations of the total system induce a mixing of K-quantum numbers. At a given angular momentum J, this configuration rotates in a *triaxial way* preferentially about the axis corresponding to the largest moment of inertia in the state with the lowest energy. Therefore the whole system rotates about the normal to the plane defined by the two pancake-like nuclei. The spins of the ^{28}Si fragments are thus in this plane since no rotation can occur about the symmetry axes of ^{28}Si. This mechanism is exactly consistent with the measured angular correlations as well as the disalignments of $L = J$. The new energy spectrum obtained after K-mixing is displayed in Fig. 2(b).

3. Decay Properties and Angular Correlations

Which molecular mode corresponds to the resonances observed? Decay properties have been investigated by means of the R-matrix theory.[4] At the resonance energy $E_{\rm cm} = 55.8$MeV, very strong excitations are seen in the single 2^+ and mutual 2^+ excitations. However the normal-mode excitations exhibit strong yields to the mutual 2^+ channel but weak enhancements to the single 2^+ channel, and thus do not fit well. Only the molecular ground state appears to follow those decay characteristics. In Fig. 3, theoretical fragment angular distributions for the molecular ground state with $J = 38$ (solid lines) are compared with the experimental data.[2] Characteristic oscillating patterns $L = J = 38$ are well reproduced. In Fig. 4, angular correlations for the mutual 2^+ excitation are compared with the data,[2] where in the panel (a), (b) and (c), the z-axis of the quantization is taken to be parallel to the incident beam, normal to the plane and the fragment

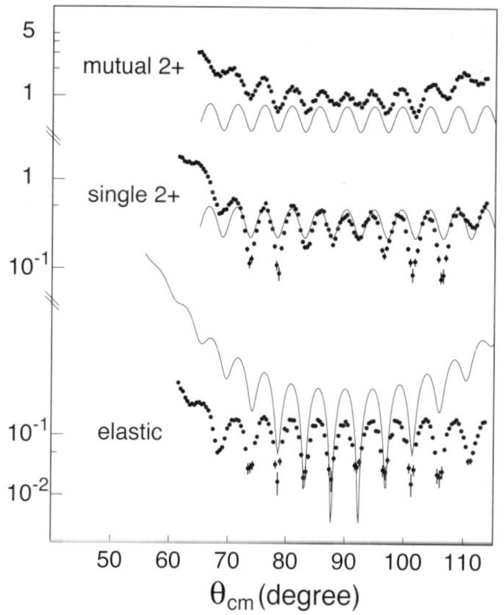

Fig. 3. Fragment angular distributions of the $^{28}Si + {}^{28}Si$ scattering.

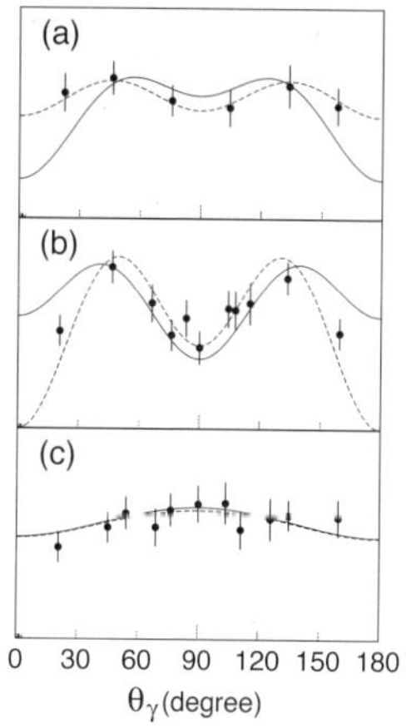

Fig. 4. Angular correlations for the mutual 2^+ excitation of $^{28}Si + {}^{28}Si$. χ^2-fits for the experimental data are given with dashed lines,[2] and theoretical results with solid lines, respectively.

direction, respectively. In Fig. 4(b), a "m=0" feature is clearly seen, and good fits are obtained in all three panels. Moreover the angular correlations for the butterfly state and the twisting state indicate clear differences, i.e., "m=2" patterns in (a) and (c) (not shown here), respectively. Thus, we can distinguish among the normal modes by the angular correlations.

4. Conclusions

The study of the $^{28}Si + {}^{28}Si$ system by the molecular model predicts a variety of states. Among them, the molecular ground state with $J = 38$ is successfully selected to be a candidate for the resonance at $E_{cm} = 55.8$ MeV. Since the angular correlations indicate different characteristics among the normal modes, we are able to identify their excitations. The experimental information on the other $^{28}Si + {}^{28}Si$ resonances is strongly called for.

References

1. R.R. Betts, B.B. Back and B.G. Glagola, *Phys. Rev. Lett.* **47**, 23 (1981)
2. R. Nouicer, C. Beck, *et al.*, *Phys. Rev. C* **60**, 041303(R) (1999).
3. E. Uegaki and Y. Abe, *Prog. Theor. Phys.* **90**, 615 (1993); *Phys. Lett. B* **340**, 143 (1994).
4. E. Uegaki and Y. Abe, *Int. J. Mod. Phys. E* **17**, 2034 (2008).

Black-sphere approximation to nuclei and its application to reactions with neutron-rich nuclei

Akihisa Kohama*

RIKEN Nishina Center, RIKEN, Saitama, Japan
**E-mail: kohama@ribf.riken.jp*

Kei Iida

Department of Natural Science, Faculty of Science, Kochi University, Kochi, Japan

Kazuhiro Oyamatsu

Department of Library and Information Science, Faculty of Letters, Aichi Shukutoku University, Aichi, Japan

We briefly review our formula for a proton-nucleus total reaction cross section, σ_R, constructed in the black-sphere approximation of nuclei, in which a nucleus is viewed as a "black" sphere of radius "a". An extension to reactions involving neutron-rich nuclei is also reported.

Keywords: Nuclear reactions, total reaction cross sections, elastic scattering, nuclear radius, neutron-rich nuclei, symmetry energy

1. Introduction and summary

Nuclear size and mass are fundamental quantities characterizing the bulk properties of nuclei. In fact, the saturation of the binding energy and density deduced from systematic data for the masses and charge radii of stable nuclei reflects the behavior of the equation of state (EoS) of nearly symmetric nuclear matter near the saturation density. The extension of such data to a more neutron-rich regime opens up an opportunity of probing the EoS of asymmetric nuclear matter.[1]

For the purpose of deducing nuclear size from proton-nucleus elastic scattering and reaction cross sections, we proposed a model in which a nucleus is viewed as a "black" (*i.e.*, strongly absorptive to incident protons) sphere of radius a, and determine a in such a way as to reproduce the angle of the observed first diffraction peak[2,3] as we will describe below. Another requirement for the black-sphere picture is that the proton wave length is considerably shorter than the nuclear size. For proton incident energies higher than about 800 MeV, these requirements are basically satisfied.

In this contribution, we briefly review our formula for a proton-nucleus total reaction cross section, σ_R, in the framework of the black-sphere approximation of nuclei, and show how to describe reactions with neutron-rich nuclei.[4] In this formula, the geometrical cross section, πa^2, is expressed as a function of the mass, A, neutron excess, $\delta \equiv (N-Z)/A$, of the target nucleus, and the kinetic energy of incident protons, T_p, in a way free from any adjustable T_p-dependent parameter. We deduce the dependence of σ_R on T_p from a simple argument involving the nuclear "optical" depth for incident protons. We describe the δ-dependence by introducing the density derivative of the symmetry energy, L, on the basis of a radius formula constructed from macroscopic nuclear models.[1,4]

We find that, for stable nuclei, this formula remarkably well reproduces the empirical T_p-dependence of σ_R at $T_p = 100 - 1000$ MeV. We show that, in this formula, the T_p-dependence of a is determined by that of nucleon-nucleon total cross sections, while the A-dependence of a is sensitive to the surface thickness of the target. For a neutron-rich case, in Fig. 1, we plot the formula values of σ_R for proton-^{A}He reactions for the cases of $L = 0$ and 100 MeV, respectively. The helium might be too light to discuss nuclear bulk properties, but, unfortunately, we have currently no data of σ_R, different from interaction cross sections, for heavier neutron-rich nuclei suitable for those studies. The fact that the formula values significantly depend on L is encouraging for future determination of L.

2. Black-sphere approximation of nuclei

In this section, we introduce the black-sphere approximation of target nuclei. The center-of-mass (c.m.) scattering angle for proton elastic scattering is generally given by $\theta_{\rm c.m.} = 2\sin^{-1}(q/2p)$ with the momentum transfer, $\mathbf{q}$, and the proton incident momentum in the c.m. frame, $\mathbf{p}$. For the proton diffraction by a circular black disk of radius a, we can calculate the value of $\theta_{\rm c.m.}$ at the first peak as a function of a. (Here we define the zeroth peak as that whose angle corresponds to $\theta_{\rm c.m.} = 0$.) We determine a in such a way that this value of $\theta_{\rm c.m.}$ agrees with the first peak angle for the measured diffraction in proton-nucleus elastic scattering, θ_M. The radius, a, and the angle, θ_M, are then related by $2pa\sin(\theta_M/2) = 5.1356\cdots$. This is obtained by requiring the derivative of the cross section with respective to the scattering angle being zero.

With the scale, a, we calculate σ_R from the present black-sphere model. Our model regards it as the geometrical cross section, $\sigma_{\rm BS} \equiv \pi a^2$. Here we assume that the incident protons are point particles, leading to vanishing contribution from the proton size to a. This is reasonable since the measured proton-proton reaction cross section is relatively small at T_p less than $\sim$ 1000 MeV. We find that the values of $\sigma_{\rm BS}$ are consistent with the measured σ_R,[3] and this feature is persistent down to $T_p \sim$ 100 MeV. Thus, the radius a can be regarded as a "reaction radius," inside which the reaction with incident protons occurs.[4] In a real nucleus, this black-sphere radius a corresponds to a radius at which the mean-free path of incident protons is of the order of the length of the penetration.

3. Black-sphere (BS)-cross-section formula

We express the energy dependence of $\sigma_{\rm R}(T_p)$ using the expression for $\sigma_{\rm BS}$ $(= \pi a^2)$ as

$$\sigma_{\rm BS}(T_p) = \pi a(T_p)^2 = \pi a_0^2 \left(1 + (\Delta a/a_0)\right)^2, \tag{1}$$

where $\Delta a \equiv a(T_p) - a_0$ and this term is responsible for the energy dependence of $\sigma_{\rm BS}$. The value of a_0 is determined at 800 MeV for each nucleus. We introduce the *effective* nuclear optical depth τ defined by $\tau = \bar{\sigma}_{pN}^{total} n_c L'$, where $\bar{\sigma}_{pN}^{total} = (Z/A)\sigma_{pp}^{total} + (1 - Z/A)\sigma_{pn}^{total}$, $\sigma_{pp(pn)}^{total}$ is the proton-proton(neutron) total cross section, n_c is the total nucleon density at the distance of $r = a$ from the nuclear center, and L' is the length of the part of the critical trajectory in which the total nucleon density is larger than n_c. By assuming that $\tau(= 0.9)$ is independent of T_p, we express the T_p-dependence of $\Delta a/a_0$ as that driven by $\bar{\sigma}_{pN}^{total}$; the T_p-*in*dependent part of $\Delta a/a_0$ is described by several parameters characterizing the density distribution of the target nucleus assumed to be trapezoidal, which makes the expression for Δa analytically tractable. The choice of the value of 0.9 for τ is reasonable since this is consistent with the values of a_0 and n_c for ^{12}C, ^{58}Ni, ^{124}Sn, and ^{208}Pb that were deduced from the elastic scattering data. The description of the formula in detail is found in the original paper.[4] The agreement of the values of the formula with the empirical values is very good for T_p higher than $\sim$ 100 MeV, where the deviation of $\bar{\sigma}_{pN}^{total}$ from its empirical value at T_p = 800 MeV is small enough to validate the present formulation. Also, the contribution from the Coulomb interaction, which is not included in this framework, can be safely neglected in this energy region.

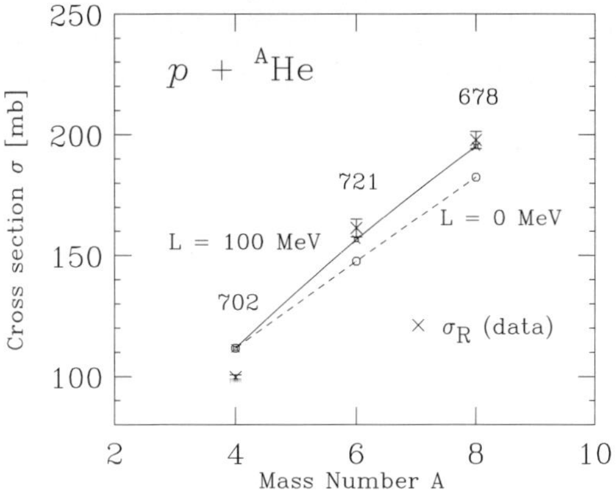

Fig. 1. Values of the BS-cross-section formula for $\sigma_{\rm R}(p + {}^A{\rm He})$ for the case of $L = 0$ (100) MeV connected by the dashed (solid) line. We adopt the values of a_0 obtained by $1.2135A^{1/3}$ [fm].[3] The empirical data[6] are plotted by ×, for which the energies of incident ^{A}He are indicated above each datum in units of MeV/nucleon.

4. An extension to neutron-rich nuclei

For the analyses of experimental data of neutron-rich unstable nuclei measured or to be measured at RI Beam Factory of RIKEN Nishina Center,[5] we could expect that the neutron-excess dependence of the black-sphere radius a would play an important role in deducing the density dependence of nuclear symmetry energy.[1] For that purpose, we introduce $f(A, \delta; L) = r_m(A, \delta; L)^2 / r_m(A, \delta = 0; L)^2$, which becomes one for $N = Z$. Here, r_m is the root-mean-square nuclear matter radius, which is obtained from macroscopic nuclear models.[1,4] Expressing $\sigma_{\rm BS}(T_p) = f(A, \delta; L)\pi a(T_p)^2$, we plot the formula values in Fig. 1.

Acknowledgments

A.K. acknowledges the organizers of this symposium for their efforts.

References

1. K. Oyamatsu and K. Iida, *Prog. Theo. Phys.* **109**, 631 (2003).
2. A. Kohama, K. Iida, and K. Oyamatsu, *Phys. Rev.* C **69**, 064316 (2004).
3. A. Kohama, K. Iida, and K. Oyamatsu, *Phys. Rev.* C **72**, 024602 (2005)
4. K. Iida, A. Kohama, and K. Oyamatsu, *J. Phys. Soc. Japan* **76**, 044201 (2007).
5. M. Takechi *et al.*, *Mod. Phys. Lett.* A **25**, 1878 (2010).
6. S. R. Neumaier *et al.*, *Nucl. Phys.* A **712**, 247 (2002).

Beyond-mean-field models for correlated nucleons. Applications of second random-phase approximation to ^{16}O and ^{48}Ca

M. Grasso*

*Institut de Physique Nucléaire, Université Paris-Sud, IN2P3-CNRS, F-91406 Orsay Cedex, France *E-mail: grasso@ipno.in2p3.fr*

D. Gambacurta and F. Catara

Dipartimento di Fisica e Astronomia and INFN, Via Santa Sofia 64, I-95123 Catania, Italy

Second RPA calculations with a Skyrme force are performed to describe both high- and low-lying excited states in ^{16}O and ^{48}Ca. The coupling between 1 particle-1 hole and 2 particle-2 hole as well as that between 2 particle-2 hole configurations are fully taken into account and the residual interaction is considered with the proper rearrangement terms. A several-MeV shift to lower energies is systematically found with respect to the RPA distributions for the strength distributions of giant resonances. The nature of the low-energy dipole excitations is analyzed in ^{48}Ca: it is found that in general these excitations cannot be in interpreted in terms of oscillations of the neutron skin against the core, with the exception of the peak with the largest $B(E1)$ value, which is located at 9.09 MeV. For this peak the neutron transition density dominates and the neutron and proton transition densities oscillate out of phase in the internal part of the nucleus leading to a strong mixing of isoscalar and isovector components. Therefore, this state shows some features usually associated to pygmy resonances.

Keywords: Beyond mean-field models, second RPA, giant resonances, pygmy modes

1. Introduction

The random-phase approximation (RPA) is currently used to describe the excitation spectrum of quantum many-body systems. Extensions and procedures to go beyond RPA and to improve the treatment of the correlations present in a many-body system have been introduced in the past decades. A natural extension of RPA is the Second RPA (SRPA) method where 2 particle-2 hole ($2p2h$) excitations are included together with the usual RPA

1 particle-1 hole ($1p1h$) configurations providing in this way a richer description of the excitation modes. Up to very recently, the SRPA equations have not been fully and self-consistently solved because the associated numerical problem is very heavy. Some approximations have been adopted in the past, like the second Tamm-Dancoff model (the matrix B is put equal to zero) or the diaghonal approximation (the equations have been solved with uncorrelated $2p2h$ states: the residual interaction terms in the matrix that couples $2p2h$ configurations among themselves have been neglected). Recently, this problem has regained a new interest; it is becoming numerically more accessible and the SRPA equations have been solved for closed-shell nuclei using an interaction derived from the Argonne V18 potential (with the Unitary Correlation Operator Method)[1,2] and for small metallic clusters in the jellium approximation.[3]

The increasing interest in the nuclear structure community is also justified by the manifestation of new phenomena in unstable nuclei. For instance, pygmy resonances represent exotic low-lying excitation modes related to the presence of a skin in neutron-rich nuclei. The necessity to go beyond standard mean field to describe these resonances has been demonstrated for example by the important effects found when the particle-vibration coupling is included.[4]

When dealing with the SRPA problem with density-dependent forces, a first formal aspect to consider is the determination of the residual interaction that has to be used in the SRPA matrix elements (many years ago a prescription has been introduced in Ref.[5]). For the first results that we have obtained in Ref.[6] for the nucleus ^{16}O we have used two currently employed prescriptions: i) we have not included any rearrangement terms in the beyond-RPA matrix elements; ii) we have calculated all the rearrangement terms with the usual RPA prescription. Important differences have been found between the two sets of results. In a recent work[7] a procedure to derive the correct expressions of all the rearrangement terms within the SRPA framework has been introduced and applied to calculations for the nucleus ^{16}O. We have employ the implemented code where the full rearrangement terms have been included to treat the low-lying excitation spectrum of the stable isotopes ^{48}Ca within the Skyrme-SRPA model.

In Sec. 2 the formal scheme of SRPA is briefly summarized. Results are presented in Sec. 3. Conclusions and perspectives are finally discussed in Sec. 4.

2. Brief summary of the formal aspects of SRPA

The excited states in SRPA are superpositions of 1 particle-1 hole ($1p1h$) and $2p2h$ configurations. The SRPA equations can be written in the compact form

$$\begin{pmatrix} \mathcal{A} & \mathcal{B} \\ -\mathcal{B}^* & -\mathcal{A}^* \end{pmatrix} \begin{pmatrix} \mathcal{X}^\nu \\ \mathcal{Y}^\nu \end{pmatrix} = \omega_\nu \begin{pmatrix} \mathcal{X}^\nu \\ \mathcal{Y}^\nu \end{pmatrix}, \qquad (1)$$

where:

$$\mathcal{A} = \begin{pmatrix} A_{11} & A_{12} \\ A_{21} & A_{22} \end{pmatrix}, \mathcal{B} = \begin{pmatrix} B_{11} & B_{12} \\ B_{21} & B_{22} \end{pmatrix},$$

$$\mathcal{X}^\nu = \begin{pmatrix} X_1^\nu \\ X_2^\nu \end{pmatrix}, \quad \mathcal{Y}^\nu = \begin{pmatrix} Y_1^\nu \\ Y_2^\nu \end{pmatrix}.$$

In the above equations, '1' and '2' stand for $1p1h$ and $2p2h$, respectively. Thus, A_{11} and B_{11} represent the usual RPA matrices, whereas the matrices A_{12} and B_{12} couple $1p1h$ with $2p2h$ configurations and the matrices A_{22} and B_{22} couple among themselves $2p2h$ configurations. If a density-dependent interaction like the Skyrme force is employed, rearrangement terms appear in the residual interaction. The usual RPA rearrangement terms appear in the matrices A_{11} and B_{11}. New types of rearrangement terms have been obtained for the other matrix elements in Ref.[7] within a variational derivation of the SRPA equations. The expressions of these rearrangement terms are reported in Ref.[7] and are used in this work.

3. Results

3.1. *The SRPA rearrangement terms:* ^{16}O

We apply the expressions for the residual interaction derived in Ref.[7] to evaluate the SRPA strength distributions associated to the monopole and quadrupole excitations in the magic nucleus ^{16}O.

We choose the same nucleus analyzed in Ref.,[8] ^{16}O. This allows us to compare the profiles obtained with the full SRPA rearrangement terms with the corresponding results obtained in Ref.[8] We use the Skyrme interaction SGII.

We present in Fig. 1 the isoscalar monopole (upper panel) and quadrupole (lower panel) strength distributions. The RPA profiles (black dashed lines) are plotted together with SRPA profiles obtained by calculating all the rearrangement terms as in RPA (SRPA$^{(a)}$), SRPA profiles obtained without rearrangement terms in beyond-RPA matrix elements (SRPA$^{(b)}$), and full SRPA profiles (red solid lines). The comparison tells us in a clear way that both the approximations which are currently used for the residual interaction in SRPA calculations lead to quite different results with respect to the full SRPA strength distributions. For the quadrupole strength, in particular, one observes that the blue dotted curve (all the rearrangement terms like in RPA) is completely flat and fragmented and this indicates that this approximation may indeed provide a very poor description of the excitation spectrum in some cases.

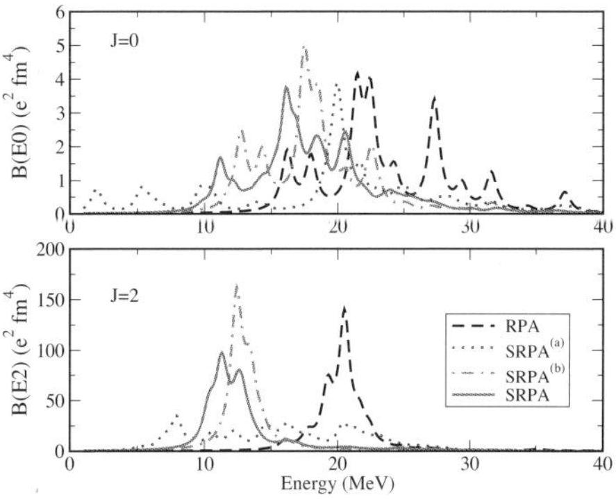

Fig. 1. Isoscalar monopole (top) and quadrupole (bottom) strength distributions obtained for ^{16}O with RPA (black dashed lines), SRPA with all the rearrangement terms as in RPA (SRPA$^{(a)}$, blue dotted lines) SRPA without rearrangement terms in beyond-RPA matrix elements (SRPA$^{(b)}$, magenta dot-dashed lines), and full SRPA (red solid lines).

As already underlined in Ref.,[8] it is found that the SRPA excitation energies are several MeV smaller than the corresponding RPA values for the giant resonances. This means that, in those cases where RPA well reproduces the experimental response of the nucleus, SRPA typically provides too low excitation energies.

3.2. *Low-lying excitations in the neutron-rich nucleus* ^{48}*Ca*

The nucleus ^{48}Ca is stable and a neutron skin has been measured experimentally by proton and electron scattering experiments.[9] The experimental low-lying dipole response has been recently analyzed[10] and a low-energy strength (between 5 and 10 MeV) has been observed.

Because of the zero-range of the Skyrme interaction, a natural energy cutoff is not provided. Different procedures to treat this problem may be envisaged for future studies (see, for instance, the exploratory work presented in Ref.[11]). In this work, we have introduced an energy cutoff ($ECUT$) on the $2p2h$ configurations. By varying it from 40 to 60 MeV we have verified that a reasonable stability of the results is achieved around a cutoff of 50-55 MeV.

The $B(E1)$ distributions for different choices of the energy cutoff are plotted in Fig. 2 up to an excitation energy of 10 MeV for the nucleus ^{48}Ca. One can observe that the results do not change strongly starting from a cutoff of 45 MeV. In what follows, we will analyze the results obtained with a cutoff of 60 MeV (bottom panel of Fig. 2).

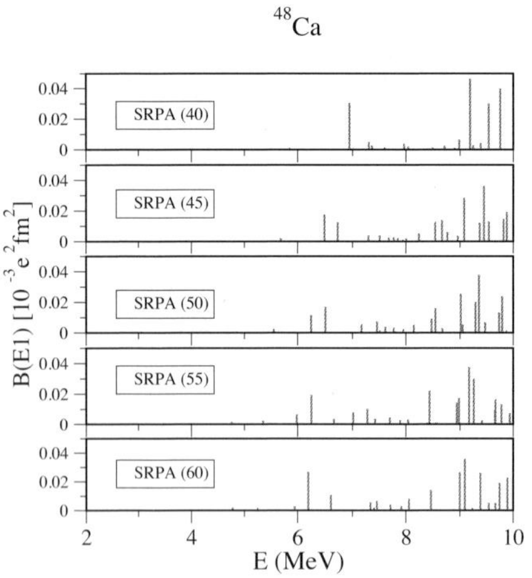

Fig. 2. (Color online) SRPA dipole strength distribution for the ^{48}Ca isotope for increasing values of the energy cutoff $ECUT$ (MeV) on the $2p2h$ configurations included in the calculations.

These results are in reasonable agreement with the experimental data and qualitatively of the same type as those found in Ref[12] with the extended theory of finite Fermi systems (ETFFS) approach which is also a beyond-mean-field model where the coupling is done with collective phonons instead of $2p2h$ configurations (as is done in SRPA).

By analyzing the transition densities we have found that in general these low-lying excitations cannot be interpreted as oscillations of the neutron skin against the core, with the exception of the peak located at 9.09 MeV. For this state the neutron transition density dominates over the proton one that is almost vanishing in the external part of the nucleus while in the interior the two densities oscillate out of phase.

4. Conclusions

We have presented Skyrme-SRPA results for the nuclei ^{16}O and ^{48}Ca. A several-MeV shift to lower energies is systematically found with respect to RPA distributions for the strength distributions of giant resonances. The nature of the low-energy dipole excitations is analyzed in ^{48}Ca: it is found that these excitations cannot be in general interpreted in terms of oscillations of the neutron skin against the core with the exception of the peak with the largest $B(E1)$ value, which is located at 9.09 MeV. For this peak the neutron transition density dominates and the neutron and proton transition densities oscillate out of phase in the internal part of the nucleus leading to a strong mixing of isoscalar and isovector components. Therefore, this state shows some features usually associated to pygmy resonances.

References

1. P. Papakonstantinou and R. Roth, Phys. Lett. **B 671**, 356 (2009).
2. P. Papakonstantinou and R. Roth, arXiv:0910.1674v1 [nucl-th].
3. D. Gambacurta and F. Catara, Phys. Rev. **B 79**, 085403 (2009).
4. G. Tertychny, V. Tselyaev, S. Kamerdzhiev, F. Krewald, J. Speth, E. Litvinova and A. Avdeenkov, Nucl. Phys. **A 788**, 159 (2007).
5. S. Adachi and S. Yoshida, Phys. Lett. **B 81**, 98 (1978).
6. D. Gambacurta, M. Grasso, and F. Catara, Phys. Rev. C 81, 054312 (2010).
7. D. Gambacurta, M. Grasso, and F. Catara, J. Phys. G: Nucl. and Part. Phys. 38, 035103 (2011).
8. D. Gambacurta, M. Grasso, and F. Catara, Phys. Rev. C 81 (2010), 054312.
9. R.H. McCamis, et al., Phys. Rev. C 33, 1624 (1986); R.F. Frosch, et al., Phys. Rev. 174, 1380 (1968).
10. T. Hartmann, J. Enders, P. Mohr, K. Vogt, S. Volz, and A. Zilges, Phys. Rev. C 65, 034301 (2002).
11. K. Moghrabi, M. Grasso, G. Colò, and N. Van Giai, Phys. Rev. Lett. 105, 262501 (2010).
12. T. Hartmann, M. Babilon, S. Kamerdzhiev, E. Litvinova, D. Savran, S. Volz, and A. Zilges, Phys. Rev. Lett. 93, 192501 (2004).

Derivation of the Interacting Boson Model from mean-field theory

Kosuke Nomura

Department of Physics, University of Tokyo,
Hongo, Bunkyo-ku, Tokyo 113-0033, Japan
E-mail: nomura@nt.phys.s.u-tokyo.ac.jp

A Hamiltonian of the interacting boson model (IBM) is determined from the mean- eld theory with nuclear energy density functionals (EDFs). The multi-nucleon dynamics of the surface deformation is simulated in terms of the boson degrees of freedom. The interaction strengths of the IBM Hamiltonian are determined so that the potential energy surfaces (PESs) of a given EDF with quadrupole degrees of freedom is compared with the corresponding PES of IBM. A fermion-to-boson mapping for a rotational nucleus is discussed. Some examples resulting from the present spectroscopic calculations are shown for neutron-rich Pt, Os and W isotopes.

Keywords: Interacting Boson Model; Density Functional Theory; shape/phase transitions;

1. Introduction

The quadrupole collectivity is a prominent feature of nuclear structure,[1] in which a nucleus can be a sphere, a deformed ellipsoid, and an object in between, depending on the number of valence nucleons. Also of interest is the evolution of the nuclear equilibrium shape at speci c nucleon numbers, which is often recognized as the quantum phase transition in nite system.[2] The understanding of these collective structural evolution from a microscopic viewpoint remains one of the major concerns in nuclear physics.[3]

The interacting boson model (IBM)[4] has been successful in phenomenology for the description of low-lying states of medium-heavy nuclei. The IBM simulates multi-nucleon dynamics of surface deformation in terms of $J = 0^+$ (s) and 2^+ (d) bosons re ecting the collective S and D pairs of valence nucleons, respectively.[5] The microscopic foundation of the IBM has been extensively studied in terms of the conventional shell-model description[5,6] as well as the newly proposed one with the energy density functional (EDF).[11] They have been practiced and shown to be valid for many cases of near spherical and -unstable shapes.

The mean- eld studies using popular EDFs e.g., of Skyrme-[7] and Gogny-types,[8] as well as those based on the relativistic mean- eld Lagrangian,[9] have been successful in the universal description of the nuclear intrinsic properties, and also have been taken as starting point for nuclear spectroscopy.[3,10,11] In Ref. 11, the potential energy surface (PES) with quadrupole degrees of freedom, calculated by mean- eld method with a given EDF, is mapped onto the corresponding PES of IBM. This produces the IBM interaction strengths, thereby allowing the spectroscopic calculations with good quantum numbers. In this report, some basic, underlying physical interpretations of the method proposed in Refs. 11,12,17 will be outlined, followed by the most recent relevant results.[12,23,24]

This contribution is based on the works done in collaboration with T. Otsuka (Tokyo), N. Shimizu (Tokyo), L. Guo (RIKEN), R. Rodríguez-Guzmán (CSIC, Madrid), L. M. Robledo (UAM, Madrid), P. Sarriguren (CSIC, Madrid), P. H. Regan (Surrey), P. D. Stevenson (Surrey), and Zs. Podolyák (Surrey).

2. Theory

We focus on the IBM-2, comprised of proton $J = 0^+$ (s) and 2^+ (d) bosons and neutron $J = 0^+$ (s) and 2^+ (d) bosons. The IBM-2 Hamiltonian to be studied is written in the form.[11,12,17]

$$H_{\mathrm{IBM}} = \epsilon(\hat{n}_{d} + \hat{n}_{d}) + \ddot{Q} \cdot \ddot{Q} + \hat{L} \cdot \hat{L}, \tag{1}$$

The rst and the second terms in Eq. (1) stand for the d-boson number operator, with ϵ denoting the single d-boson energy relative to the s boson one, and the quadrupole-quadrupole interaction, respectively. The third term is the rotational-kinetic like term.

The nucleonic PES is obtained from the constrained Hartree-Fock+BCS (HF+BCS) or Hartree-Fock-Bogoliubov (HFB) calculation. Then, a point on the mean- eld PES is mapped onto the corresponding point on the PES of IBM. Here, the PES of boson system is an expectation value of $\hat{H}_{\mathrm{IBM}}$ in terms of the boson coherent state, $| \ \rangle \propto \prod_{\rho= \ ,} \left[s_\rho^\dagger + \sum_{\mu=0,\pm2} a_{\rho\mu} d^\dagger_{\rho\mu} \right]^{n_\rho} |0\rangle$, where the coe cients $a_{\rho\mu}$ correspond to the quadrupole deformation variables and .[1] n_ρ and $|0\rangle$ stand for the number of proton or neutron bosons and an inert core, respectively. The geometrical shape of a given nucleus is described by these variables. The interaction strengths of the IBM Hamiltonian of Eq. (1) are determined by drawing the IBM PES so that it reproduces the overall pattern of the mean- eld PES.[12] Note that the $L \cdot L$ term is not determined from the PES because it contributes to the PES in the same way as the $\hat{n}_{d\rho}$ term, and is thus neglected in this step. The pattern of the PES re ects the e ects of the basic (fermion) features of many-nucleon system such as nuclear forces and the anti-symmetrization. By the mapping of the PES, these important fermion properties should be simulated in terms of bosonic degrees of freedom.

Here, the total energy of mean- eld calculation is compared with the corresponding total energy of the boson system. Therefore, both vibrational and rotational kinetic energies, similar to those introduced when solving a ve-dimensional collective Hamiltonian from a given EDF,[14–16] are supposed to be included in the boson system. One should, however, go one step beyond in the case of strong deformation, where the nucleon intrinsic wave function di ers notably from the corresponding boson intrinsic wave function.[17] The di erence results in the too small IBM rotational moment of inertia compared to the experimental one.[11,12] The deviation of the moment of inertia occurs probably because the degrees of freedom are limited for boson system, which may require other types of bosons such as those with spin 4 (g boson). In Ref. 17, we have proposed an alternative which considers the mapping of the rotational response, i.e., cranking, without changing the intrinsic wave function. The $L \cdot L$ term in Eq. (1) makes most straightforward contribution to the rotational moment of inertia and is introduced so that the rotational response of IBM becomes similar to that of the nucleon system. The $L \cdot L$ coupling constant is determined by the comparison of the cranking moments of inertia between fermion and boson systems, as illustrated in detail in Ref. 17.

Taking all these things into account, the low-lying states of Sm isotopes are calculated as shown in Fig. 1, where fermionic PES is obtained by the constrained HF+BCS approach with Skyrme SkM*[19] in a usual manner.[3,10] Levels are calculated by NPBOS code.[18] The agreement of the calculated yrast levels with experimental ones[20] are rather well particularly for deformed $N \geqslant 90$ Sm nuclei. The rapid decrease of the yrast spectra, as well as the kinks observed in 0_2^+ and 2_2^+ energies, re ect the rst-order shape/phase transition[2] with critical point at $N = 90$.[21] The present result follows this trend, although the transition occurs less sharply. Nevertheless, it is worth while stressing that there is no phenomenological adjustment in calculating the spectra.

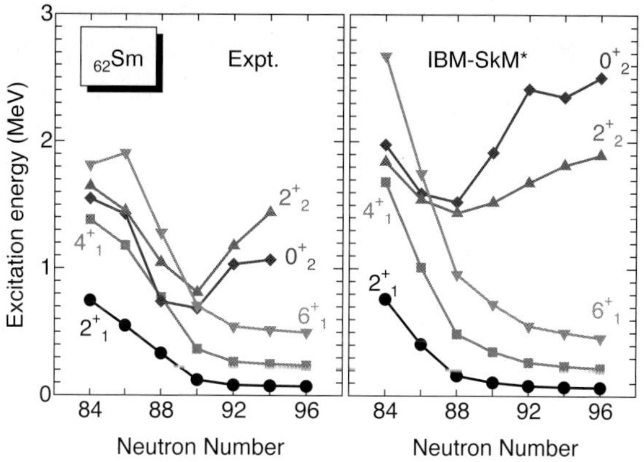

Fig. 1. Theoretical low-lying states of Sm isotopes as compared to the experiment.[20] Skyrme SkM* force is used.

3. Applications to Pt, Os and W isotopes with Gogny EDF

Figure 2 shows the evolution of low-lying (a) ground-state (g.s.), (b) quasi- and (c) quasi- band energies of the Pt isotopes. The IBM parameters are determined based on the Hartree-Fock-Bogoliubov calculations with the Gogny-D1S EDF.[25] The $L \cdot L$ term is neglected here, as the problem as to the scale of the IBM rotational spectra does not occur. The theoretical spectra in Fig. 2 agree with the experimental ones.

In Fig. 2(a) there are clear ngerprints for structural evolution with a jump between ^{186}Pt and ^{188}Pt, which can be correlated with the change of the PESs from prolate to oblate deformations.[23] In addition, in Fig. 2, the 0^+_2 and the 2^+_3 levels look nearly constant from A =180 to 186 in both theory and experiment. These two levels are pushed up rather signi cantly from A=186 to 188 In the experimental quasi- -band spectra in Fig. 2(c), the 3^+_1 level lies close to the 4^+_2 level for $180 \leqslant A \leqslant 186$. This trend persists in the present calculation even for $188 \leqslant A \leqslant 196$, suggesting the feature characteristic of O(6) symmetry where the staggering occurs as 2^+_γ, $(3^+_\gamma\ 4^+_\gamma)$, $(5^+_\gamma\ 6^+_\gamma)$, ... etc. However, the experimental levels are lying more regularly. The deviation is a consequence of an algebraic nature of the IBM. The so-called cubic interaction[26] may change the situation. This should be investigated further in the context of the re nement of the boson Hamiltonian.

Figures 2(d) and 2(e) show the g.s.-band and the quasi- -bandhead 2^+_γ energies for Os and W isotopes, respectively. What is of interest is the be-

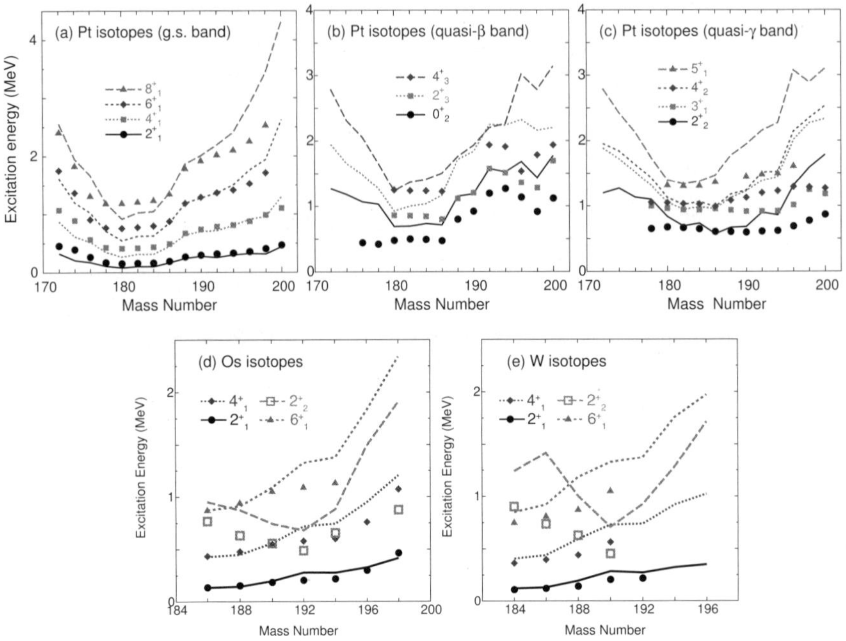

Fig. 2. Low-lying states of (a)(b)(c) Pt, (d) Os and (e) W isotopes. Gogny D1S EDF is used. Experimental data[20] are identi ed as symbols.

havior of 2^+_γ energy, exhibiting a kink for ^{192}Os and ^{190}W. The experimental 2^+_γ energy in ^{192}Os is lower than the 4^+_1 one. Thus the $A = 192$ nucleus is the most -unstable one among Os isotopes. A similar trend is predicted for W isotopes, exhibiting more rapid change as a function of N. The location of the 2^+_γ state for ^{196}Os (^{192}W) has not yet been xed experimentally, but the 4^+_1 state is lower than the 2^+_γ one in both ^{196}Os and ^{192}W in the calculation.

4. Conclusions

The recently proposed approach to determine the IBM Hamiltonian from a modern EDF was illustrated, together with the application to W, Os, and Pt isotopes. A fermion-boson mapping for rotational deformed nuclei was outlined. With these approaches, spectroscopic properties can be computed with good quantum numbers in the laboratory frame. More spectroscopic predictions on the heavy exotic nuclei will be expected, which should be examined experimentally in the near future.

Acknowledgments

This work is supported in part by grants-in-aid for scienti c research (A) 20244022 and No.217368, and by JSPS Core-to-Core program EFES.

References

1. A. Bohr and B. R. Mottelson, *Nuclear Structure*, Vol.II (Benjamin, New York, 1975).
2. P. Cejnar, R. F. Casten, and J. Jolie, Rev. Mod. Phys. **82**, 2155 (2010).
3. P. Ring and P. Schuck, *The Nuclear Many-Body Problem* (Springer, Berlin, 1980).
4. F. Iachello and A. Arima, *The interacting boson model* (Cambridge University Press, Cambridge, 1987)
5. T. Otsuka, A. Arima, and F. Iachello, Nucl. Phys. A **309**, 1 (1978).
6. T. Mizusaki and T. Otsuka, Prog. Theor. Phys., Suppl. **125**, 97 (1997)
7. T. H. R. Skyrme, Nucl. Phys. **9**, 615 (1959) ; D. Vautherin and D. M. Brink, Phys. Rev. C **5**, 626 (1972).
8. J. Decharge, M. Girod, and D. Gogny, Phys. Lett. B **55**, 361 (1975).
9. D. Vretenar *et al.*, Phys. Rep. **409**, 101 (2005).
10. M. Bender, P.-H. Heenen, P.-G. Reinhard, Rev. Mod. Phys. **75**, 121 (2003).
11. K. Nomura, N. Shimizu, and T. Otsuka, Phys. Rev. Lett. **101**, 142501 (2008)
12. K. Nomura, N. Shimizu, and T. Otsuka, Phys. Rev. C **81**, 044307 (2010)
13. A. E. L. Dieperink *et al.*, Phys. Rev. Lett. **44**, 1747 (1980).
14. P. Bonche *et al.*, Nucl. Phys. A **510**, 466 (1990).
15. J.-P. Delaroche *et al.*, Phys. Rev. C **81**, 014303 (2010).
16. Z. P. Li *et al.*, Phys. Rev. C **81**, 034316 (2010).
17. K. Nomura, T. Otsuka, N. Shimizu, and L. Guo, to appear in Phys. Rev. C Rapid Communications, arXiv:1011.1056.
18. T. Otsuka and N. Yoshida, JAERI-M report **85** (Japan At. Enr. Res. Inst., 1985)
19. J. Bartel *et al.*, Nucl. Phys. A **386**, 79 (1982).
20. National Nuclear Date Center, http://www.nndc.bnl.gov/nudat2/.
21. F. Iachello, Phys. Rev. Lett. **87**, 052501 (2001)
22. E. Chabanat *et al.*, Nucl. Phys. A **635**, 231 (1998).
23. K. Nomura, T. Otsuka, R. Rodríguez-Guzmán, L. M. Robledo, and P. Sarriguren, Phys. Rev. C **83**, 014309 (2011).
24. K. Nomura, T. Otsuka, R. Rodríguez-Guzmán, L. M. Robledo, P. Sarriguren, P. H. Regan, P. D. Stevenson, and Zs. Podolyák, *submitted* [arXiv:1101.1699].
25. J. F. Berger, M. Girod, D. Gogny, Nucl. Phys. A **428** (1984) 23c.
26. K. Heyde *et al.*, Phys. Rev. C **29** 1420 (1984).

Large-amplitude quadrupole collective dynamics in neutron-rich Mg and Cr isotopes

N. HINIOHARA[*,1], K. SATO[1], K. YOSHIDA[2,1], T. NAKATSUKASA[1], M. MATSUO[2] and K. MATSUYANAGI[1,3]

[1]*RIKEN Nishina Center, Wako 351-0198, Japan*
[2]*Department of Physics, Faculty of Science, Niigata University, Niigata 950-2181, Japan*
[3]*Yukawa Institute for Theoretical Physics, Kyoto University, Kyoto 606-8502, Japan*
[*]*E-mail: hinohara@riken.jp*

The miroscopic method of deriving the quadrupole collective Hamiltonian is presented. Shape changes and transitions in neutron-rich magnesium and chromium isotopes are discussed using the quadrupole collective Hamiltonian.

Keywords: Large-amplitude collective motion; Shape phase transition

1. Introduction

Spectroscopic studies of low-lying excited states are revealing many interesting and novel phenomena in unstable nuclei. To understand their properties, theoretical analysis using the collective Hamiltonian derived from nuclear energy density functional is a powerful tool, with which we can calculate experimental observables, such as energy spectra and electromagnetic transition probabilities. In addition, the collective wave functions provide us with an intuitive collective nature of the individual nuclear state. The quantities in the five-dimensional collective Hamiltonian which should be determined in a microscopic way are collective potential, three vibrational masses, and three rotational moments of inertia (inertial functions). The inertial functions are usually calculated within the Inglis-Belyaev (IB) cranking approximation. The most serious shortcoming of the IB cranking approximation is that the time-odd terms of the moving mean field are ignored. To overcome this shortcoming, we have recently formulated the constrained Hartree-Fock-Bogoliubov plus local quasiparticle random-phase approximation (CHFB + LQRPA) method[1] to calculate the collective inertial functions on the basis of the adiabatic self-consistent collective coordinate (ASCC) method.[2] This new method enables us to evaluate the inertial functions including the time-odd contribution of the moving mean The field. It has been successfully applied to the oblate-prolate shape coexistence/mixing phenomena in proton-rich Se and Kr isotopes.[1,3]

In this presentation the large-amplitude deformation dynamics in low-lying states of neutron-rich magnesium and chromium isotopes is discussed with the use of the CHFB+LQRPA method.

2. CHFB + LQRPA method

The five-dimensional quadrupole collective Hamiltonian[4] is constructed with the CHFB+LQRPA method. In the CHFB + LQRPA method, the collective potential is calculated from the CHFB states $|\phi(\beta,\gamma)\rangle$, which are determined by solving the CHFB equations with four constraints on neutron and proton numbers and quadrupole deformations (β,γ). The vibrational and rotational inertial functions are derived from the LQRPA equations. They are the QRPA equations on top of the CHFB states $|\phi(\beta,\gamma)\rangle$. See Ref. 1 for more detailed description of the theory.

3. Application to the neutron-rich Mg isotopes

We derive the five-dimensional quadrupole collective Hamiltonian for the large-amplitude shape dynamics in neutron-rich magnesium isotopes around ^{32}Mg. The significant increases of the $E(4_1^+)/E(2_1^+)$ ratio and the $B(E2; 2_1^+ \to 0_1^+)$ value from ^{30}Mg to ^{34}Mg clearly indicate a rapid growth of quadrupole deformation associated with the breaking of $N = 20$ shell gap. The experimental data suggest a kind of quantum phase transition taking place around ^{32}Mg and stimulate a microscopic investigation on large-amplitude collective dynamics unique to this region of the nuclear chart.

Numerical calculations are performed using the pairing-plus-quadrupole (P+Q) Hamiltonian including the quadrupole-pairing interaction within the two-major harmonic-oscillator shell model space. The parameters in the model Hamiltonian are determined to reproduce the results of the HFB calculation using the Skyrme density functional and surface-type pairing. See Ref. 5 for more detailed description of the model parameters.

The potential energy surface clearly indicates that the prolate deformation grows up with increasing neutron number. The potential energy surface for ^{30}Mg is very soft against the axial quadrupole deformation, while that for ^{32}Mg exhibits a spherical and prolate shape coexistence, where the spherical shape is associated with the neutron shell gap at $N = 20$. The spherical local minima disappears in the collective potentials of ^{34,36}Mg, and the prolate minima become soft in the triaxial direction.

Figure 1 shows the excitation energies and the $B(E2)$ values obtained by requantizing the collective Hamiltonian. The calculation reproduces the experimental trend, such as the lowering of the excitation energy of 2_1^+ state and the increase of the $E(4_1^+)/E(2_1^+)$ ratio from ^{30}Mg to ^{34}Mg. More detailed analysis including the properties of the excited bands is now in progress.

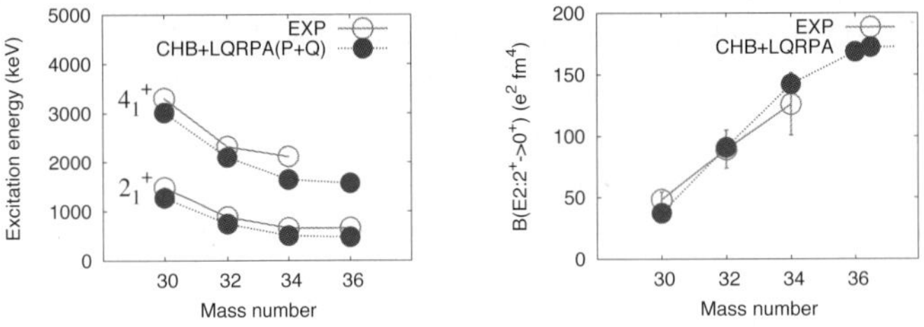

Fig. 1. Excitation energies and $B(E2)$ for yrast states of magnesium isotopes. Figure is taken from Ref. 5.

4. Derivation of collective Hamiltonian from density functional

We discuss the direct derivation of the collective Hamiltonian from the Skyrme density functional. We start from the (1+2)-dimensional version of the collective Hamiltonian, where the axial quadrupole vibration along the β direction and rotations about the two axes which are perpendicular to the symmetry axis are taken into account.[6] We solve the CHFB + LQRPA equations employing the SkM* Skyrme energy density functional in the particle-hole channel, and the volume-type pairing interaction in the pairing channel.[7] See Ref. 6 for more detailed description of the theory.

We investigate the properties of large-amplitude collective dynamics from ^{58}Cr to ^{68}Cr, where the development of deformation toward $N = 40$ is indicated from the observed small excitation energy of 2_1^+ state.

The collective potentials of ^{58,60}Cr show softness against the β deformation. We even get the HFB minima at $\beta = 0$. Beyond ^{62}Cr, the prolate minimum gradually develops up to ^{68}Cr. We found a shoulder or a shallow minimum in the oblately deformed region as well.

We successfully determine the one vibrational and two rotational moments of inertia by solving the LQRPA equations, and obtain the excitation energies and $B(E2)$ values in the yrast states. When the neutron number increases from ^{58}Cr, $E(2_1^+)$ drops toward ^{62}Cr. The $E(4_1^+)/E(2_1^+)$ ratio and the transition probability $B(E2; 0_1^+ \to 2_1^+)$ grow up at the same time. These features are consistent with the experimental results, and they clearly show the onset of deformation at ^{62}Cr.

5. Conclusion

We have determined microscopically the quadrupole collective Hamiltonian with the CHFB + LQRPA method and investigated the shape phase transition in the low-lying states of neutron-rich magnesium and chromium isotopes. The full five-dimensional quadrupole collective Hamiltonian is microscopically derived in magnesium isotopes. The excitation energies and $B(E2)$ are in good agreement with the experimental data which indicates the development of the prolate deformation, and the shape phase transition in this region. For the chromium isotopes, the quadrupole collective Hamiltonian is directly derived from the nuclear energy density functional, and the shape phase transition is discussed. The calculation gives consistent results with the experimental data for the properties of the low-lying excited states. The onset of the deformation takes place at ^{62}Cr.

Acknowledgments

N. H. is supported by the Special Postdoctoral Researcher Program of RIKEN. The numerical calculations were carried out on RIKEN Cluster of Clusters (RICC) facility and T2K-Tsukuba. This work is supported by KAKENHI (Nos. 21340073 and 20105003).

References

1. N. Hinohara, *et al.*, *Phys. Rev. C* **82**, 064313 (2010).
2. M. Matsuo, *et al.*, *Prog. Theor. Phys.* **103**, 959 (2000).
3. K. Sato and N. Hinohara, *Nucl. Phys. A* **849**, 53 (2011).
4. L. Próchniak and S. G. Rohoziński, *J. Phys. G* **36**, 123101 (2009).
5. N. Hinohara, *et al.*, *AIP Conf. Proc.* **1355**, 200 (2011).
6. K. Yoshida and N. Hinohara, *Phys. Rev. C* **83**, 061302 (2011).
7. H. Ôba and M. Matsuo, *Prog. Theor. Phys.* **120**, 143 (2008).

Microscopic description of large-amplitude collective motions with local QRPA inertial masses

K. SATO[*,1], N. HINOHARA[1], T. NAKATSUKASA[1],
M. MATSUO[2], K. MATSUYANAGI[1,3]
[1] *RIKEN Nishina Center, Wako 351-0198, Japan*
[*] *E-mail: satok@ribf.riken.jp*

[2] *Department of Physics, Faculty of Science, Niigata University, Niigata 950-2181, Japan*

[3] *Yukawa Institute for Theoretical Physics, Kyoto University, Kyoto 606-8502, Japan*

We propose a practical method for determining the collective potential and inertial functions in the five-dimensional quadrupole collective Hamiltonian. We have applied this method to large-amplitude shape mixing dynamics in the low-lying states of proton-rich Kr isotopes.

Keywords: Large-amplitude collective motion; shape coexistence.

1. Introduction

The five-dimensional (5D) quadrupole collective Hamiltonian has been used as a standard tool for the description of quadrupole collective motions in the low-lying states in atomic nuclei. It is written in terms of the magnitude of quadrupole deformation β, the degree of triaxiality γ, the three Euler angles and their time derivatives. The 5D quadrupole collective Hamiltonian is characterized by seven quantities, i.e., the collective potential and three vibrational and three rotational inertial functions, and they are in general functions of deformation (β, γ). To calculate the inertial functions, the so-called Inglis-Belyaev (IB) cranking formula has been used widely. However, it is known that the IB cranking formula overestimates the experimental excitation energies because it does not contain the contribution from the time-odd components of the mean field.

Recently, we have developed a method for determining microscopically the inertial functions in the collective Hamiltonian, with which the contribution from the time-odd components is taken into account.[1,2] The method is called the constrained Hartree-Fock-Bogoliubov (CHFB) plus local quasiparticle RPA (LQRPA) method after that we solve the LQRPA equations, which are an extension of the usual QRPA equations to the non-HFB equilibrium points, on top of each CHFB state in the (β, γ) plane.

Here we shall show some results of the application of the CHFB+LQRPA method to the low-lying states in proton-rich Kr isotopes, where oblate-prolate shape coexistence has been conjectured experimentally.

2. Theoretical framework

The 5D quadrupole collective Hamiltonian is written as

$$\mathcal{H}_{\rm coll} = T_{\rm vib} + T_{\rm rot} + V(\beta, \gamma), \tag{1}$$

$$T_{\rm vib} = \frac{1}{2} D_{\beta\beta}(\beta, \gamma)\dot{\beta}^2 + D_{\beta}\ \ (\beta, \gamma)\dot{\beta}\dot{\gamma} + \frac{1}{2} D\ \ (\beta, \gamma)\dot{\gamma}^2, \tag{2}$$

$$T_{\rm rot} = \frac{1}{2} \sum_{k=1}^{3} \mathcal{J}_k(\beta, \gamma)\omega_k^2, \tag{3}$$

where $T_{\rm vib}$, $T_{\rm rot}$ and V represent the vibrational, rotational and collective potential energies, respectively.

In the CHFB+LQRPA method, we first solve the CHFB equation,

$$\delta \langle \phi(\beta, \gamma) | \hat{H}_{\rm CHFB}(\beta, \gamma) | \phi(\beta, \gamma) \rangle = 0, \tag{4}$$

to determine the collective potential $V(\beta, \gamma)$, imposing the constraints on the particle numbers and quadrupole deformation parameters (β, γ). Then, we solve the LQRPA equations,

$$\delta \langle \phi(\beta, \gamma) | [\hat{H}_{\rm CHFB}(\beta, \gamma), \hat{Q}^i(\beta, \gamma)] - \frac{1}{i}\hat{P}_i(\beta, \gamma) | \phi(\beta, \gamma) \rangle = 0, \tag{5}$$

$$\delta \langle \phi(\beta, \gamma) | [\hat{H}_{\rm CHFB}(\beta, \gamma), \frac{1}{i}\hat{P}_i(\beta, \gamma)] - C_i(\beta, \gamma)\hat{Q}^i(\beta, \gamma) | \phi(\beta, \gamma) \rangle = 0, \tag{6}$$

on top of the CHFB states obtained above. The vibrational inertial functions are calculated by transforming two LQRPA modes to the (β, γ) degrees of freedom. We also solve the LQRPA equations for rotation to determine the moments of inertia. Finally, we solve the collective Schrödinger equation to obtain the excitation energies, collective wave functions, $E2$ transition strengths and moments.

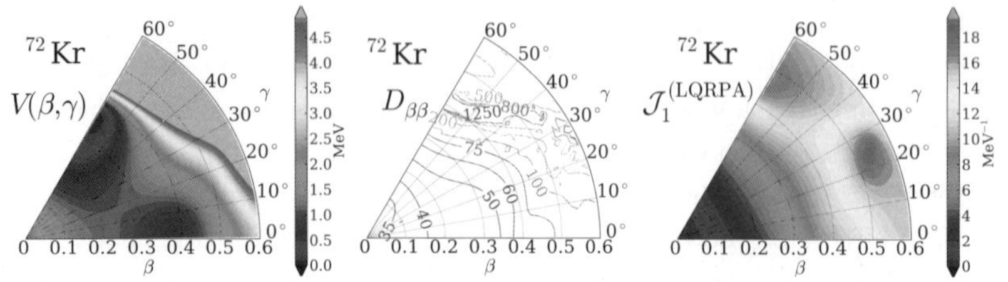

Fig. 1. Collective potential $V(\beta, \gamma)$ in units of MeV (*left*), LQRPA vibrational inertial mass $D_{\beta\beta}(\beta, \gamma)$ (*middle*) and LQRPA rotational moment of inertia $\mathcal{J}_1(\beta, \gamma)$ (*right*) in units of MeV^{-1} calculated for ^{72}Kr.

3. Numerical Results

Here, we shall show the results of the application of the CHFB+LQRPA method to ^{72}Kr.[2] Figure 1 shows the collective potential $V(\beta, \gamma)$, the vibrational inertial function $D_{\beta\beta}(\beta, \gamma)$ and the rotational moment of inertia $\mathcal{J}_1(\beta, \gamma)$ for ^{72}Kr. The collective potential has two local minima.Reflecting this potential, the collective wave function localizes on the oblate side in the ground band, whereas it localizes on the prolate side in the excited band due to the orthogonality to the yrast states. In fact, the degree of shape mixing is strongly related to the $\beta - \gamma$ dependence of the moments of inertia as well as the collective potential. We have seen that the asymmetry of the moments of inertia between the oblate and prolate sides causes the development of the localization of the collective wave function in the (β, γ) plane as angular momentum increases.[2]

Figure 2 shows the excitation energies and $B(E2)$ values calculated with the CHFB+LQRPA method in comparison with those calculated using the IB cranking inertial functions. The excitation energies evaluated with the LQRPA masses are lower than those calculated with the IB cranking masses and are in better agreement with the experimental data. Note that the interband transition between the initial and final states having the same angular momentum becomes weaker with increasing the angular momentum. This results from the development of the localization of the collective wave functions mentioned above and indicates that the shape-coexistence-like character becomes stronger with increasing angular momentum.

4. Conclusion

We have determined microscopically the five-dimensional quadrupole collective Hamiltonian with the CHFB+LQRPA method and investigated the

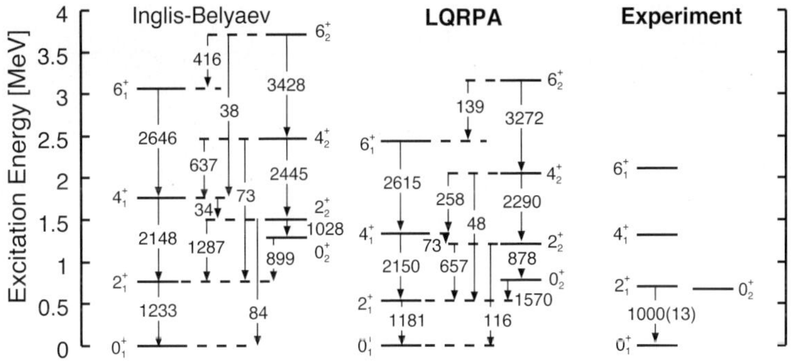

Fig. 2. Excitation spectra and $B(E2)$ values calculated for ^{72}Kr by means of the CHFB+LQRPA method (denoted LQRPA) and experimental data.[3–5] For comparison, results calculated using the IB cranking masses (denoted Inglis-Belyaev) are also shown. Only $B(E2)$'s larger than 1 Weisskopf unit are shown in units of $e^2\text{fm}^4$.

shape coexistence/mixing phenomena in the low-lying state of proton-rich Kr isotopes. The excitation energies calculated with the LQRPA inertial functions are in better agreement with the experimental data than those obtained with the widely-used IB cranking masses. Numerical results also suggest that the effect of rotational motion plays an important role in shape mixing dynamics.

Acknowledgments

N. H. is supported by the Special Postdoctoral Researcher Program of RIKEN. The numerical calculations were carried out on Altix3700 BX2 at Yukawa Institute for Theoretical Physics in Kyoto University and RIKEN Cluster of Clusters (RICC) facility. This work is supported by the Core-to-Core Program of the Japan Society for the Promotion of Science "International Research Network for Exotic Femto Systems".

References

1. N. Hinohara, K. Sato, T. Nakatsukasa, M. Matsuo and K. Matsuyanagi, *Phys. Rev. C* **82**, p. 064313 (2010).
2. K. Sato and N. Hinohara, *Nucl. Phys. A* **849**, p. 53 (2011).
3. E. Bouchez et al., *Phys. Rev. Lett.* **90**, p. 082502 (2003).
4. A. Gade et al., *Phys. Rev. Lett.* **95**, p. 022502 (2005).
5. S. M. Fischer et al., *Phys. Rev. C* **67**, p. 064318 (2003).

TWO-PROTON RADIOACTIVITY AS A TOOL FOR NUCLEAR STRUCTURE

B. Blank, P. Ascher, L. Audirac, J. Giovinazzo, N. Adimi, G. Canchel, F. Delalee, C.E. Demonchy, C. Dossat, S. Grévy, L. Hay, J. Huikari, T. Kurtukian-Nieto, S. Leblanc, I. Matea, J.-L. Pedroza, J. Pibernat, L. Serani

Centre d'Études Nucléaires de Bordeaux Gradignan - Université Bordeaux 1 - UMR 5797 CNRS/IN2P3, Chemin du Solarium, BP 120, 33175 Gradignan, France

F. de Oliveira Santos, L. Perrot, P.C. Srivastava, C. Stodel, J.-C. Thomas

Grand Accélérateur National d'Ions Lourds, CEA/DSM - CNRS/IN2P3, Bd. Becquerel, BP 55027, 14076 Caen Cedex 05, France

C. Borcea, I. Companis

National Institute for Physics and Nuclear Engineering, P.O. Box MG6, Bucharest-Margurele, Romania

B. A. Brown

Department of Physics and Astronomy, and National Superconducting Cyclotron Laboratory, Michigan State University, East Lansing, Michigan 48824-1321, USA

L. V. Grigorenko

Flerov Laboratory of Nuclear Reactions, JINR, Dubna 141980, Russian Federation

The decay of the ground-state two-proton emitters ^{45}Fe and ^{54}Zn was studied with a time-projection chamber and the emission of two protons was unambiguously identified. The total decay energy and the half-life measured agree with the results from previous experiments. The energy sharing between the two protons and their relative emission angle was determined with a time projection chamber and is compared to model predictions. In parallel, we observed directly β-delayed two- and three-proton emission from ^{43}Cr, a known β-delayed two-proton emitter.

Keywords: exotic nuclei, two-proton radioactivity, time-projection chamber

1. Introduction

Goldanskii[1] predicted, for the most proton-rich nuclei with an even number of protons, the occurance of a new nuclear decay mode, which he termed two-proton (2p) radioactivity. He suggested that, due to the pairing of protons, nuclei exist which are bound with respect to one-proton emission but are unbound to two-proton emission. It is interesting that the list of possible 2p emitters proposed by Goldanskii contained already ^{45}Fe, for which ground-state two-proton radioactivity was indeed discovered.[2,3]

After the observation of this decay mode for ^{45}Fe, it was demonstrated also in ^{54}Zn[4] and possibly in ^{48}Ni.[5] The later evidence was recently confirmed by an experiment at MSU.[6] In these discovery experiments, the ions of interest were deeply implanted in silicon detectors. Therefore, no individual protons could be observed and only the total decay energy, the half-life, and the absence of β particles from the competing decay by β-delayed-particle emission could be clearly established. In addition, the observation of the daughter decay helped to unambiguously assign the observed decay to 2p radioactivity. These experimental results were found in good agreement with predictions from different theoretical models,[7] like the R-Matrix theory,[8] the Shell Model embedded in the Continuum (SMEC)[9] or the three-body model.[10,11]

A gaseous detector, a time projection chamber (TPC) was necessary to observe the two protons individually and to measure their energy and their relative angle. This was achieved at CENBG.[12] Here we present results from experiments with the TPC to study ^{45}Fe, ^{54}Zn, and ^{43}Cr.

2. Decay of ^{45}Fe

^{45}Fe isotopes were produced by means of projectile fragmentation of a ^{58}Ni beam at 75 MeV/nucleon and were selected by the LISE3 separator and directed to the TPC installed at the exit of the LISE3 beam line. Silicon detectors allowed the fragments to be identified on an event-by-event basis. The isotopes of interest were stopped in the center of the TPC and their decay could be studied.[13]

Figure 1 shows decay events correlated to a ^{45}Fe implantation as reconstructed from the signals in energy and time registered by the different strips. These spectra allow us to determine the energy of each individual proton and the angle between the two protons. These data are shown in figure 2. We use two different presentations of these results: i) each event is represented at the angle determined by the analysis and ii) each event is dis-

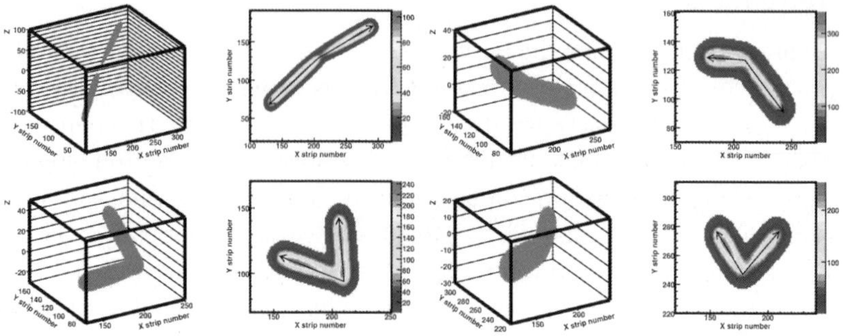

Fig. 1. Four ^{45}Fe decay events are presented in two and three dimensions as reconstructed from the energy signals of the strips alone and from the energy and time signals, respectively.

tributed in angle according to the precision obtained for the angle, i.e. the larger the error on the angle the larger the Gaussian over which the event is distributed. The lines show the distributions as predicted from the three-body model of Grigorenko *et al.*[10] Although our statistics is not enough to formulate definitive statements, it is evident that our distributions are in agreement with the theoretical predictions.

These data will allow us to extract nuclear structure information by means of a quantitative comparison with theoretical models like the nuclear shell model and the three-body model of Grigorenko.[11] The work is under way for ^{45}Fe.

The present results are in agreement with the data obtained at MSU.[14]

3. Decay of ^{54}Zn

Before the ^{54}Zn experiment, the TPC has been modified in order to cure some of the deficiencies of the first prototype. In particular, an entrance at 45° with respect to the strip directions has been chosen in order to avoid saturation effects of strips parallel to the beam direction. Also we worked at slightly higher pressure (750 mbar instead of 500 mbar) to stop more efficiently the ions of interested.[15]

The 2D and 3D spectra of ^{54}Zn ressemble those of ^{45}Fe and they are not presented here. The energy sharing between the two protons and the angle between them are shown in figure 3. As in the case of ^{45}Fe, the protons share equally the available energy. From the angular distribution, we could determine that the two protons have a p^2 composition of the wave function in ^{54}Zn of 30^{+33}_{-21}%. When we add to this spectroscopic factors from

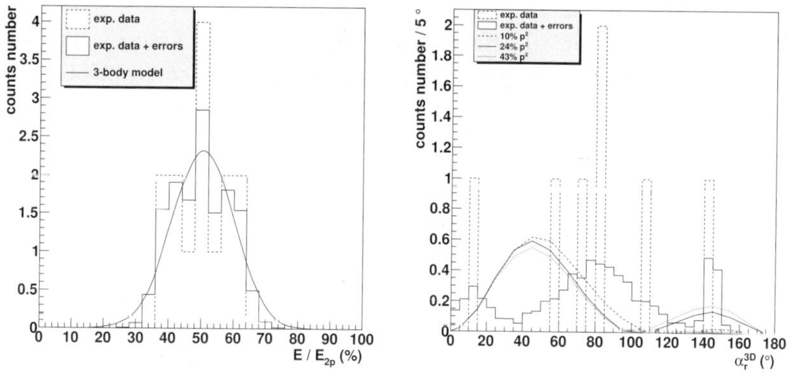

Fig. 2. Left: Spectrum of the energy correlation of the two protons emitted by ^{45}Fe. The solid histogram shows experimental data convoluted with error bars and the dashed histogram shows the raw data. The curve allows to compare data to predictions of the three-body model of Grigorenko. Right: Angular distribution of the two protons. The curves correspond to calculations with the three-body model for different wave function compositions.

shell-model calculations and the emission dynamics as obtained from the three-body model, we can determine that 99% of the 2p emission goes via this p^2 part of the wave function.

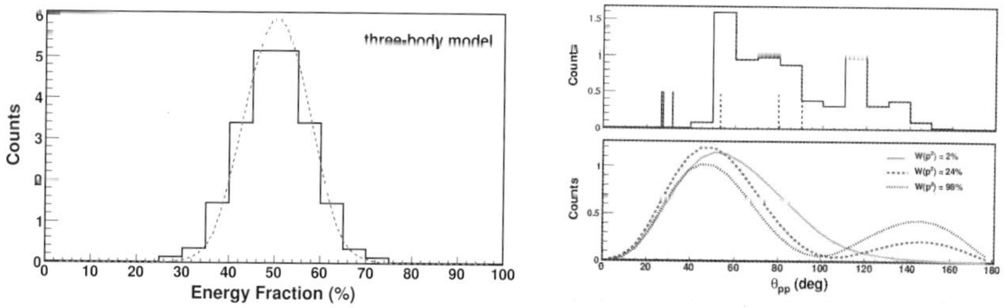

Fig. 3. Left: Spectrum of the energy correlation of the two protons emitted by ^{54}Zn. The solid histograms show the experimental data convoluted with error bars. The line is a prediction from the three-body model of Grigorenko. Right: Same for the angular distribution of the two protons. The bars are events not fully reconstructed for which two angles are possible.

4. Beta-delayed three-proton emission of ^{43}Cr

During the experiment to study the decay of ^{45}Fe, ^{43}Cr was also implanted in the TPC. Two events of β-delayed 3p emission could be observed (figure 4). The branching ratios have been determined to be 87.1(25)%, 12.7(10)%, and $0.14^{+0.19}_{-0.09}$% for the 1p, 2p, and 3p branches, respectively.

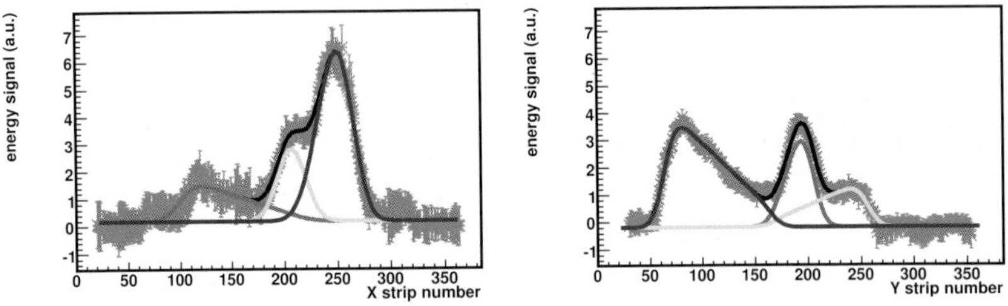

Fig. 4. One of the two β-delayed 3p events observed in the decay of ^{43}Cr. The spectra for the X and Y strip directions are shown.

5. Conclusion

The measurements persented here demonstrate the power of TPCs to study multi-particle decays. In particular, two-proton radioactivity has been studied in the cases of ^{45}Fe and of ^{54}Zn and a β3p branch was observed in the decay of ^{43}Cr. Future experiments will aim at higher statistics and will try to find additional 2p emitters to obtain a larger picture of this new and rather exotic decay mode.

References

1. V. I. Goldansky, *Nucl. Phys.* **19**, p. 482 (1960).
2. J. Giovinazzo, B. Blank, M. Chartier, S. Czajkowski, A. Fleury, M. L. Jimenez, M. Pravikoff, J.-C. Thomas, F. de Oliveira Santos, M. Lewitowicz, V. Maslov, M. Stanioiu, R. Grzywacz, M. Pfützner, C. Borcea and B. Brown, *Phys. Rev. Lett.* **89**, p. 102501 (2002).
3. M. Pfützner, E. Badura, C. Bingham, B. Blank, M. Chartier, H. Geissel, J. Giovinazzo, L. Grigorenko, R. Grzywacz, M. Hellström, Z. Janas, J. Kurcewicz, A. Lalleman, C. Mazzocchi, I. Mukha, G. Münzenberg, C. Plettner, E. Roeckl, K. Rykaczewski, K. Schmidt, R. Simon, M. Stanoiu and J.-C. Thomas, *Eur. Phys. J.* **A14**, p. 279 (2002).
4. B. Blank, A. Bey, G. Canchel, C. Dossat, A. Fleury, J. Giovinazzo, I. Matea, N. Adimi, F. D. Oliveira, I. Stefan, G. Georgiev, S. Grévy, J. C. Thomas,

C. Borcea, D. Cortina, M. Caamano, M.Stanoiu, F. Aksouh, B. A. Brown, F. Barker and W. A. Richter, *Phys. Rev. Lett.* **94**, p. 232501 (2005).

5. C. Dossat, A. Bey, B. Blank, G. Canchel, A. Fleury, J. Giovinazzo, I. Matea, N. Adimi, F. D. Oliveira, I. Stefan, G. Georgiev, S. Grévy, J. C. Thomas, C. Borcea, D. Cortina, M. Caamano, M.Stanoiu, F. Aksouh, B. A. Brown, F. Barker and W. A. Richter, *Phys. Rev. C* **72**, p. 054315 (2005).
6. M. Pomorski, M. Pfützner, W. Dominik, R. Grzywacz, T. Baumann, J. S. Berryman, H. Czyrkowski, R. Dabrowski, T. Ginter, J. Johnson, G. Kaminski, A. Kuzniak, N. Larson, S. N. Liddick, M. Madurga, C. Mazzocchi, S. Mianowski, K. Miernik, D. Miller, S. Paulauskas, J. Pereira, K. P. Rykaczewski, A. Stolz and S. Suchyta, *Phys. Rev. C* **83**, p. 061303 (2011).
7. B. Blank and M. Płoszajczak, *Rev. Prog. Phys.* **71**, p. 046301 (2008).
8. B. A. Brown and F. C. Barker, *Phys. Rev. C* **67**, p. 041304 (2003).
9. J. Rotureau, J. Okolowicz and M. Ploszajczak, *Nucl. Phys.* **A767**, p. 13 (2006).
10. L. V. Grigorenko, B. V. Danilin, V. D. Efros, N. B. Shulgina and M. V. Zhukov, *Phys. Rev. C* **60**, p. 044312 (1999).
11. L. Grigorenko, R. Johnson, I. Thompson and M. Zhukov, *Phys. Rev. C* **65**, p. 044612 (2002).
12. B. Blank, L. Hay, J. Huikari, S. Leblanc, S. List, J.-L. Pedroza, P. Ascher, L. Audirac, C. Borcea, G. Canchel, F. Delalee, C. Demonchy, C. Dossat, J. Giovinazzo, P. Hellmuth, C. Marchand, I. Matea, R. de Oliveira, J. Pibernat, A. Rebii, L. Serani and J. C. Thomas, *Nucl. Instrum. Meth.* **A613**, p. 65 (2010).
13. L. Audirac *et al.*, *to be published* (2011).
14. K. Miernik, W. Dominik, Z. Janas, M. Pfützner, L. Grigorenko, C. R. Bingham, H. Czyrkowski, M. Ćwiok, I. G. Darby, R. Dabrowski, T. Ginter, R. Grzywacz, M. Karny, A. Korgul, W. Kuśmierz, S. N. Liddick, M. Rajabali, K. Rykaczewski and A. Stolz, *Phys. Rev. Lett.* **99**, p. 192501 (2007).
15. P. Ascher, L. Audirac, N. Adimi, B. Blank, C. Borcea, B. A. Brown, I. Companis, F. Delalee, C. Demonchy, F. de Oliveira Santos, J. Giovinazzo, S. Grévy, L. V. Grigorenko, T. Kurtukian-Nieto, S. Leblanc, J.-L. Pedroza, L. Perrot, J. Pibernat, L. Serani, P. Srivastava and J.-C. Thomas, *accepted for publication in PRL* (2011).

^{14}Be$(P, N)^{14}$B REACTION AT 69 MEV AND PROPORTIONALITY RELATIONSHIP BETWEEN FORWARD ANGLE (P, N) CROSS SECTION AND B(GT)

Y. SATOU

Department of Physics and Astronomy, Seoul National University, Seoul 151-742, Korea

T. NAKAMURA, Y. KONDO, N. MATSUI, Y. HASHIMOTO, T. NAKABAYASHI, T. OKUMURA, M. SHINOHARA

Department of Physics, Tokyo Institute of Technology, Tokyo 152-8551, Japan

N. FUKUDA, T. SUGIMOTO, H. OTSU, Y. TOGANO, T. MOTOBAYASHI, H. SAKURAI, Y. YANAGISAWA, N. AOI, S. TAKEUCHI, T. GOMI, M. ISHIHARA

RIKEN Nishina Center, Saitama 351-0198, Japan

S. KAWAI

Department of Physics, Rikkyo University, Tokyo 171-8501, Japan

H. J. ONG, T. K. ONISHI

Department of Physics, University of Tokyo, Tokyo 113-0033, Japan

S. SHIMOURA, M. TAMAKI

Center for Nuclear Study (CNS), University of Tokyo, Saitama 351-0198, Japan

T. KOBAYASHI, Y. MATSUDA, N. ENDO, M. KITAYAMA

Department of Physics, Tohoku University, Miyagi 980-8578, Japan

Using the proportionality relationship between the forward angle (p, n) cross section and the reduced transition probability $B(\mathrm{GT})$, the $|J_{\sigma\tau}(q=0)|$ values have been deduced for recent (p, n) reaction data on ^{14}Be at E_p=69.4 MeV and for existing data on ^{12}C at various incident energies. A discussion was made on the energy dependence of $|J_{\sigma\tau}(q=0)|$ and on mutual consistency between the values for ^{14}Be and ^{12}C.

Keywords: Gamow-Teller transition; (p, n) reaction; DWBA.

1. Introduction

The (p, n) charge-exchange (CE) reaction in inverse kinematics is an important tool for studies of exotic nuclear structure of nuclei far from β stability. The Gamow-Teller (GT) transitions are strongly excited in the CE reaction at small momentum transfer q. The GT transitions involving unstable nuclei are of great interest not only for understanding spin-isospin correlations in nuclei,[1] but also for their implications for a variety of astrophysical phenomena where weak processes play a key role.[2] Being unconstrained by Q_β limitations, the nuclear CE reaction has a unique capability to be exploited as a practical source of information of GT strengths at a high excitation energy region of exotic nuclei. Recently we have reported an experimental study of the (p, n) reaction on a neutron drip-line nucleus ^{14}Be, performed within the invariant mass method, leading to the particle unbound 1^+ state at 1.27 MeV in ^{14}B.[3]

In this note a further discussion was made using the data on the proportionality relationship between the forward angle (p, n) cross section and the reduced transition probability $B(\mathrm{GT})$. Such a relationship is of fundamental importance in relating the observable with nuclear structure information. An empirical value for the volume integral of the central spin-isospin part of the nucleon-nucleon (NN) interaction at q=0, $|J_{\sigma\tau}(q=0)|$, was extracted and compared with those deduced from a similar analysis of existing ^{12}C$(p, n)^{12}$N$(1^+, \mathrm{g.s.})$ reaction data at various incident energies.

2. Proportionality Relationship

A factorized expression for the (p, n) cross section at q=0 is given by[4,5]

$$(d\sigma/d\Omega)_{q=0}^{L=0} = \left[\frac{E_i E_f}{(\hbar^2 c^2 \pi)^2}\right] N_D |J_{\sigma\tau}|^2 B(\mathrm{GT}). \tag{1}$$

Here $E_{i(f)}$ is the reduced energy in the initial (final) channel. The distortion factor N_D is the ratio of distorted-wave and plane-wave cross sections at q=0. The cross section $(d\sigma/d\Omega)_{q=0}^{L=0}$, defining the L=0 contribution at q=0, is obtained by extrapolating the 0° cross section to q=0, with an additional factor given by the ratio of calculated cross sections obtained with only the central terms of the NN interaction and with the full force components.

In Fig. 1 (a), forward angle (p, n) cross sections involving GT transitions are shown as a function of q for the ^{14}Be$(p, n)^{14}$B$(1^+, 1.27\ \mathrm{MeV})$ reaction at E_p=69.4 MeV[3] and the ^{12}C$(p, n)^{12}$N$(1^+, \mathrm{g.s.})$ reactions at E_p=61.8 MeV,[6] 120 MeV,[7] 135 MeV,[8] 186 MeV,[9] and 296 MeV.[10] Solid lines are the results

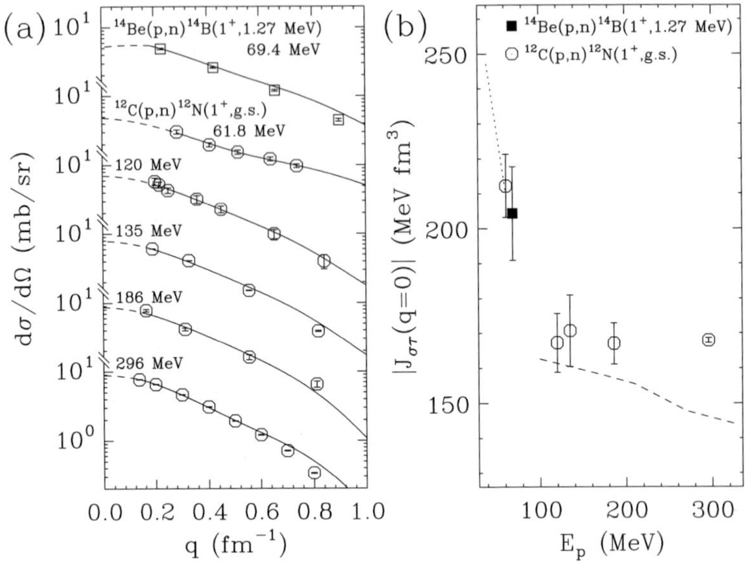

Fig. 1. (a) Forward angle (p, n) cross sections for the $^{14}Be(p, n)^{14}B(1^+, 1.27$ MeV) and $^{12}C(p, n)^{12}N(1^+, g.s.)$ reactions in comparison with DWBA calculations (solid and dashed lines). (b) Values for the volume integral $|J_{\sigma\tau}|$ extracted using Eq. (1).

of microscopic distorted-wave Born approximation (DWBA) calculations normalized to the data. In the calculations the optical model potential was taken from the semi-microscopic parametrization of JLMB.[11] The effective NN interaction was the M3Y interaction[12] below 100 MeV, while it was taken to be the t-matrix interaction of Franey and Love[13] above 100 MeV. The one-body transition density was calculated using the shell model within the $0\hbar\omega$ configuration. The WBT and P(10-16)T Hamiltonians[14] were respectively used for ^{14}Be and ^{12}C. The single particle wave functions were generated in a harmonic oscillator well with oscillator parameters of b=2.04 and 1.87 fm for ^{14}Be and ^{12}C, respectively.[3] We see that the angular distributions are well described by the present DWBA framework. Momentum extrapolation to q=0 was made by changing the Q value. The extrapolated cross sections are shown as dashed lines.

With known B(GT) values for the $^{14}Be(\beta^-)^{14}B(1^+, 1.27$ MeV) transition: B(GT)=0.80(9),[15] and for the inverse of the $^{12}N(1^+, g.s.)(\beta^+)$ $^{12}C(0^+, g.s.)$ transition: B(GT) =0.875(6),[16] the values for the volume integral $|J_{\sigma\tau}|$ were deduced using Eq. (1). The results are plotted in Fig. 1 (b). The error bars are quadratic sums of statistical and systematic uncertainties. The dashed and dotted lines are theoretical predictions of Refs. 13 and 17, respectively. It is seen that the $|J_{\sigma\tau}|$ values do not show any significant energy dependence above 100 MeV. The extraction of $B(\mathrm{GT})$ from the (p,n) reaction measurements would thus be more favorably performed above 100 MeV. The $|J_{\sigma\tau}|$ value for the $^{14}\mathrm{Be}(\beta^-)^{14}\mathrm{B}(1^+, 1.27\ \mathrm{MeV})$ transition is found to be consistent with those derived from the inverse of the $^{12}\mathrm{N}(1^+, \mathrm{g.s.})(\beta^+)^{12}\mathrm{C}(0^+, \mathrm{g.s.})$ transition. It would be interesting to see if there exist universal (energy-dependent) $|J_{\sigma\tau}|$ values in the mass region of current interest in forthcoming studies.

3. Summary

Energy dependent $|J_{\sigma\tau}|$ values were deduced from an analysis of the $^{12}\mathrm{C}(p,n)^{12}\mathrm{N}(1^+, \mathrm{g.s.})$ cross sections at various incident energies. The energy dependence of $|J_{\sigma\tau}|$ was observed to be weak above 100 MeV. The $|J_{\sigma\tau}|$ value extracted from the $^{14}\mathrm{Be}(p,n)^{14}\mathrm{B}(1^+, 1.27\ \mathrm{MeV})$ reaction data was found to be consistent with those from the carbon target data.

Acknowledgments

The work was in part supported by the NRF grants (Nos. R32-2008-000-10155-0 (WCU), 2010-0027136) of MEST Korea.

References

1. M.N. Harakeh, A. van der Woude, *Giant Resonances* (Oxford University Press, New York, 2001).
2. K. Langanke, G. Martínez-Pinedo, *Rev. Mod. Phys.* **75**, 819 (2003)
3. Y. Satou *et al.*, *Phys. Lett. B* **697**, 459 (2011).
4. C.D. Goodman *et al.*, *Phys. Rev. Lett.* **44**, 1755 (1980).
5. T.N. Taddeucci *et al.*, *Nucl. Phys. A* **469**, 125 (1987).
6. B.D. Anderson *et al.*, *Nucl. Instrum. Methods* **169**, 153 (1980).
7. J. Rapaport *et al.*, *Phys. Rev. C* **24**, 335 (1981).
8. B.D. Anderson *et al.*, *Phys. Rev. C* **54**, 237 (1996).
9. X. Yang *et al.*, *Phys. Rev. C* **52**, 2535 (1995).
10. T. Wakasa *et al.*, *Phys. Lett. B* **656**, 38 (2007).
11. E. Bauge, J.P. Delaroche, M. Girod, *Phys. Rev. C* **63**, 024607 (2001).
12. G. Bertsch *et al.*, *Nucl. Phys. A* **284**, 399 (1977).
13. M.A. Franey and W.G. Love, *Phys. Rev. C* **31**, 488 (1985).
14. E.K. Warburton, B.A. Brown, *Phys. Rev. C* **46**, 923 (1992).
15. N. Aoi *et al.*, *Phys. Rev. C* **66**, 014301 (2002).
16. F. Ajzenberg-Selove, *Nucl. Phys. A* **506**, 1 (1990).
17. W.G. Love, in *The (p,n) Reaction and the Nucleon-Nucleon Force*, eds. C.D. Goodman *et al.*, (Plenum, New York, 1980), pp. 23.

EXPLORING Mg ISOTOPE STRUCTURES THROUGH BETA-DELAYED DECAY OF SPIN-POLARIZED Na ISOTOPES

T. Shimoda[1,*], K. Tajiri[1], K. Kura[1], A. Odahara[1], T. Hori[1], M. Kazato[1], T. Masue[1], M. Suga[1], A. Takashima[1], T. Suzuki[1,†], T. Fukuchi[1,‡], Y. Hirayama[2], N. Imai[2], H. Miyatake[2], M. Pearson[3], C.D.P. Levy[3], K.P. Jackson[3]

[1]*Department of Physics, Osaka University, Osaka 560-0043, Japan*
[2]*High Energy Accelerator Research Organization (KEK), Ibaraki 305-0801, Japan*
[3]*TRIUMF, 4001 Wesbrook Mall, Vancouver, BC, V6T 2A3, Canada*

The present work aims at experimentally clarifying the structure change of neutron-rich Mg isotopes as a function of the neutron number by firmly establishing the level schemes of the Mg isotopes in a wide excitation energy range. We use our unique method of β-decay spectroscopy which takes advantage of anisotropic β decay of spin-polarized Na isotopes. The asymmetry parameters of the β decay are the effective tool to unambiguously assign the spin-parity of levels in the daughter nucleus. As the first step of the systematic studies on neutron-rich Mg isotopes from lighter region outside of the $N = 20$ "island of inversion" to heavier one across through the island, very detailed level scheme of ^{28}Mg and ^{29}Mg have been successfully established at TRIUMF, where the world-highest polarized alkali beams are available.

Keywords: polarized RI beam, β-delayed decay spectroscopy, ^{28}Mg and ^{29}Mg, island of inversion, intruder configurations, shell evolution

1. Introduction

One of the recent highlights in nuclear physics is the shell evolution in very neutron-rich nuclei with variation of neutron number. In particular, the exotic structure of nuclei around the "island of inversion",[1] the very neutron-rich region in a nuclear chart with the neutron numbers close to the magic number $N = 20$, has been attracting much attention for two decades. The significant fractions of intruder configurations in the ground and/or excited

*shimoda@phys.sci.osaka-u.ac.jp
†present address : RCNP, Osaka University, Japan
‡present address : RIKEN Center for Molecular Imaging Science, Kobe, Japan

states should cause very different features from those of the magic-number nuclei close to the stability line. There have been many experimental and theoretical works to clarify which nucleus locates on the island of inversion and which level has structure with intruder configurations. However, the experimental information on these nuclei are rather limited. For many of the levels, spin and parity, which are the key quantities to understand the nuclear structure, have not been assigned. For example, in the case of ^{29}Mg, only the ground-state spin-parity has been assigned.[2,3]

We have applied an effective method of β-decay spectroscopy to investigate the structure of Mg isotopes. This method takes advantage of asymmetric β-decay of the spin-polarized Na to assign spin-parity of the levels of daughter nucleus Mg. The present paper briefly discusses the successful results on ^{28}Mg and ^{29}Mg. The detail will be found elsewhere.[4,5]

2. Experiment

The principle of the method is based on the asymmetric β-decay shown by the following angular dependence, if we assume allowed transition:

$$W(\theta) \simeq 1 + AP\cos\theta, \tag{1}$$

where P, θ and A are the polarization of the parent nucleus, the emission angle of the β-ray with respect to the polarization direction and the asymmetry parameter characteristic of β-transition, respectively. The essential point is that the asymmetry parameter A is a constant depending on the initial and final state spins, I_i and I_f, respectively, as shown in Eq.(2) and Table 1. The large difference in A enables us to unambiguously assign I_f.

$$A \begin{cases} = -1 & (I_f = I_i - 1), \\ \simeq \dfrac{-1}{I_i+1} & (I_f = I_i), \\ = \dfrac{I_i}{I_i+1} & (I_f = I_i + 1). \end{cases} \tag{2}$$

Table 1: Asymmetry parameters for allowed β-decay of ^{28}Na and ^{29}Na.

β-decay	I_i^π	I_f^π	A
		0^+	-1.0
$^{28}\mathrm{Na}_{\mathrm{g.s.}} \to {}^{28}\mathrm{Mg}$	1^+	1^+	-0.5
		2^+	$+0.5$
		$1/2^+$	-1.0
$^{29}\mathrm{Na}_{\mathrm{g.s.}} \to {}^{29}\mathrm{Mg}$	$3/2^+$	$3/2^+$	-0.4
		$5/2^+$	$+0.6$

The experiments were performed at TRIUMF, where the world-highest polarization is available for radioactive alkali beams.[6] Figure 1 shows the layout of detectors. The 30.4 keV polarized $^{28}Na^+$ or $^{29}Na^+$ beam was stopped on a thin Pt foil in vacuum. The effective polarization of ^{28}Na and ^{29}Na was 28.3(5)% and 35(2)%, respectively. The β-rays and the successive γ-rays were detected by 9 detectors surrounding the Pt foil. Two detectors ("Left" and "Right") placed on the polarization axis measured the asymmetry of β-decay.

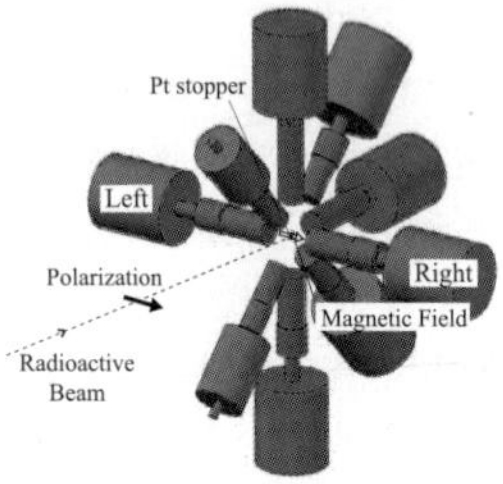

Fig. 1. Schematic detector arrangement.

3. Results

In ^{28}Mg two levels at high excitation energy were newly found and the overall good agreement of the level scheme with the shell-model calculations in the *sd*-shell model space indicates that the structure of ^{28}Mg is understood mostly based on the normal configurations.

Figure 2 shows the decay scheme of ^{29}Na revised by the present work. The physical quantities and decay paths marked by asterisks are newly found ones in the present work. It is seen that two new γ-transitions have been found and spins and parities (I^π) of 7 levels have been assigned in ^{29}Mg for the first time. Note that the spins and parities of the most of ^{29}Mg states except for the two levels at 1.094 and 1.430 MeV were assigned. These levels are characterized by large log ft-values.

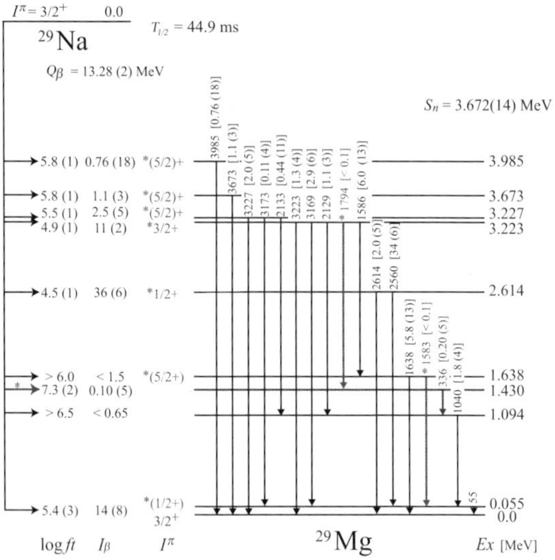

Fig. 2. Revised decay scheme of ^{29}Na.

The experimental results were compared with the shell-model calculations with the USDB Hamiltonian.[7] It was found that the level energy, spin-parity and β-transition probability were well reproduced assuming all nucleons are in *sd*-shell for all the positive-parity levels. However, the spin-unassigned two levels could not be explained by such calculations. This fact strongly suggests that these levels at 1.094 and 1.430 MeV are the negative-parity levels with intruder configuration. This result is consistent with the previous speculation.[8] In fact, the Monte Carlo Shell Model calculation taking into account the *pf*-shell configurations predicts $7/2^-$ and $3/2^-$ levels at 0.68 MeV and 1.01 MeV, respectively.[9]

Now, it becomes possible to compare the excitation energies of the negative-parity levels in neutron rich Mg isotopes. Figure 3 shows the negative-parity levels in odd-mass Mg isotopes, assuming $7/2^-$ and $3/2^-$ assignments for the 1.430- and 1.094-MeV ^{29}Mg levels, respectively. It is clearly seen that the excitation energies of the negative-parity levels rapidly decrease at ^{29}Mg, when the neutron number increases. This is the experimental evidence for the shell evolution in neutron-rich Mg isotopes.

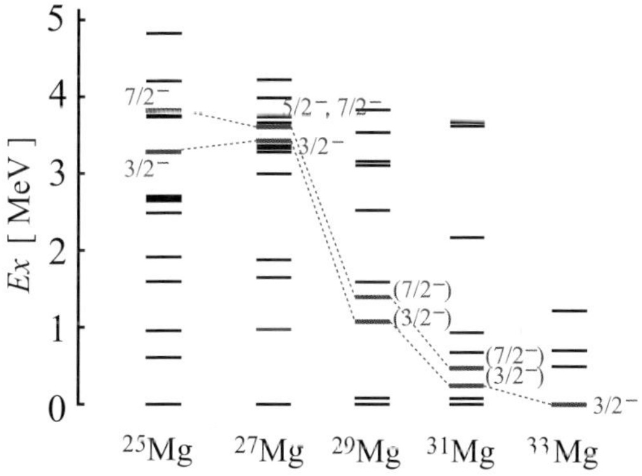

Fig. 3. Low-lying negative-parity levels in odd-mass Mg isotopes. The spin-parity assignments of the ^{31}Mg levels and $^{33}Mg_{g.s.}$ are after Ref.[8] and Refs.,[10] respectively

References

1. E. K. Warburton, J. A. Becker and B. A. Brown, Phys. Rev. C **41**, 1147 (1990).
2. P. Baumann *et al.*, Phys. Rev. C **36**, 765 (1987).
3. National Nuclear Data Center, Brookhaven National Laboratory, [http://www.nndc.bnl.gov/].
4. K. Kura *et al.*, submitted to Phys. Rev. C (2012).
5. K. Tajiri, PhD thesis, Osaka University (2012).
6. C. D. P. Levy *et al.*, Nucl. Instrum. Methods B **204**, 689 (2003).
7. B. A. Brown and W. A. Richter, Phys. Rev. C **74** (2006) 034315.
8. J. R. Terry *et al.*, Phys. Rev. C **77** (2008) 014316.
9. Y. Utsuno: private communication.
10. D. T. Yordanov *et al.*, Phys. Rev. Lett. **99** (2007) 212501; Hyperfine Interact. **196** (2010) 53.

A NEW METHOD TO EXPLORE HIGH-SPIN STATES BY RI BEAM INDUCED FUSION REACTION

A. Odahara[1,*], T. Shimoda[1], Y. Ito[1], H. Nishibata[1], K. Tajiri[1], J. Takatsu[1], N. Hamatani[1 †], R. Yokoyama[1 ‡], C. Petrache[2,3], R. Leguillon[2], T. Suzuki[4], E. Ideguchi[5], H. Watanabe[6], Y. Wakabayashi[7], K. Yoshinaga[8], D. Beaumel[3], P. Desesquelles[9], D. Curien[10], D. Guinet[11], G. Lehaut[11]

[1] *Department of Physics, Osaka University, Osaka 560-0043, Japan*
[2] *Department of Physics, University of Paris Sud XI, Orsay F-91406, France*
[3] *Institut de Physique Nucleaire d'Orsay (IPNO), Orsay F-91406, France*
[4] *Research Center for Nuclear Physics (RCNP), Osaka University, Osaka 567-0047, Japan*
[5] *Center for Nuclear Study (CNS), University of Tokyo, Saitama 351-0198, Japan*
[6] *RIKEN, Saitama 351-0198, Japan*
[7] *Japan Atomic Energy Agency (JAEA), Tokai-mura, Ibaraki 319-1195, Japan*
[8] *Department of Physics, Tokyo University of Science, Chiba 278-8510, Japan*
[9] *Centre de Spectrometrie Nucleaire et de Spectrometrie de Masse (CSNSM), Orsay F-91406, France*
[10] *Institut Pluridisciplinaire Hubert Curien (IPHC), Strasbourg 67037, France*
[11] *Institut de Physique Nucleaire de Lyon (IPNL), Villeurbanne 69622, France*

A new γ-ray spectroscopy method was developed using low-energy (around 5-10 MeV/u) RI beam induced fusion reaction, which enables us to study high-spin states of nuclei in wider mass region where high-spin states could not be produced by the combination of the stable beam and stable target. The low-energy ^{17}N RI beam was produced by the direct reaction using low-energy primary beam and was transported up to the secondary target using the EN beam line at RCNP, Osaka University. Two experiments were performed to investigate the high-spin states in ^{142}Pr (N=83 isotone) and in ^{136}Ba (N=80 isotone) by this new γ-ray spectroscopy method using the low-energy ^{17}N RI beam.

Keywords: RI beam induced fusion reaction, low-energy RI beam, high-spin states, γ-ray spectroscopy

*odahara@phys.sci.osaka-u.ac.jp
†present address : RCNP, Osaka University, Osaka 567-0047, Japan
‡present address : CNS, University of Tokyo, Saitama 351-0198, Japan

1. Low-energy RI beam induced fusion reaction

High-spin states have been studied in the limited mass region, such as the neutron deficient nuclei close to the β-stability line, by fusion reaction using the combination of the stable beam and stable target. New γ-ray spectroscopy method using fusion reaction induced by the low-energy (around 5-10 MeV/u) RI beam enables us to search for the exotic deformations, collective and single-particle motions of unobserved high-spin states in wider mass region. This low-energy RI beam can be directly produced from the low-energy stable beam by in-flight method. This new γ-ray spectroscopy method using fusion reaction induced by the RI beam has an advantage for the high-S/N measurement, although RI beam with low-beam intensity has a disadvantage in statistics. The high-S/N improvement can be made by the ways as follows: (1) By using the RI beam, the best combination of the RI beam and the stable target with the largest cross section and ratio to total can be selected for the fusion reaction. (2) The relatively low-beam intensity enables the event-by-event detection of the γ rays correlated with the RI beam.

2. Experimental procedures and results

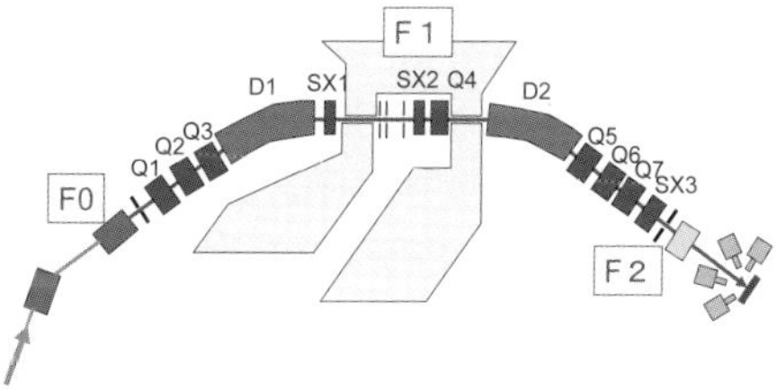

Fig. 1. RCNP secondary beam line.

New γ-ray spectroscopy method using fusion reaction induced by the direct low-energy RI beam was applied to the experiments at Research Center for Nuclear Physics (RCNP), Osaka University. Low-energy ^{17}N RI beam was delivered using the RCNP secondary beam line, EN beam line,[1,2] as shown in Fig. 1. This beam line consists of 2 dipole, 7 quadrupole, 3 sextupole magnets with 2 focal planes of dispersive at F1 and achromatic at F2. As the rooms installing the primary target, F1 and F2 are separated by concrete wall each other, γ rays can be measured at secondary target position of F2 under low-background condition with no background originating from the primary beam with high-beam intensity. Two kinds of experiments were performed to study the high-spin states in ^{142}Pr (N=83 isotone) and in ^{136}Ba (N=80 isotone).

2.1. *Production of ^{17}N RI beam*

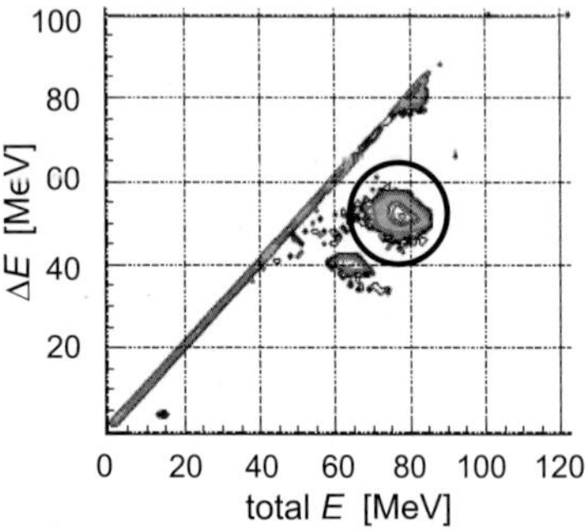

Fig. 2. Particle identification of secondary beams at F2.

Nucleus ^{17}N was produced by the direct reaction of ^{9}Be(^{18}O,^{17}N)^{10}B. Primary $^{18}O^{5+}$ beam of 9.2 and 9.3 MeV/u provided by the azimuthally-varying-field (AVF) cyclotron bombarded a ^{9}Be target of 10 mg/cm^2 with an average intensity of 0.8 and 1.5 pμA to investigate high-spin states in ^{142}Pr and ^{136}Ba, respectively. The ^{18}O beam was deflected slightly around 1.8 degree at the primary target so as to make the primary beam stop as far as possible from the γ-ray measurement system. The magnetic rigidity selection was optimized for $^{17}N^{7+}$ with an acceptance of $\Delta(B\rho)/B\rho$=2 % at F1. Aluminum foils of around 50 and 44 μm were set at F1 as the uniform degrader in 2 experiments for ^{142}Pr and ^{136}Ba, respectively. Particle identification of RI beams was performed using 2 Si detectors with thickness of 72 and 250 μm installed at F2 to make the ΔE-E spectrum, as shown in Fig. 2. In this figure, $^{17}N^{7+}$ is indicated to be encircled. Purity of the ^{17}N RI beam at F2 was around 60 % by the mixture of mainly $^{18}O^{8+}$and $^{19}O^{6+}$ particles. The 4.9 and 5.5 MeV/u ^{17}N RI beam with the beam intensity of 2×10^4 and 1.2×10^5 pps was transported up to F2. The ^{17}N RI beam was focused into a spot size of around 8 mm in diameter on the secondary target.

2.2. *Study for high-spin states in ^{142}Pr*

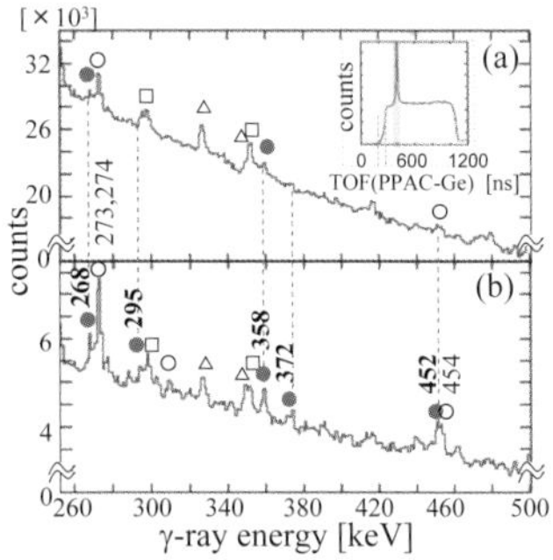

Fig. 3. Gamma-ray singles spectra. The inset shows the spectrum of TOF(PPAC-Ge).

The experiment was performed to study high-spin states in ^{142}Pr, one of N=83 isotones, which was produced by the ^{130}Te(^{17}N,5n) secondary fusion reaction. Gamma rays were measured by 14 high-purity Ge detectors surrounding the secondary target at a distance of 100-140 mm. The total detection efficiency was obtained to be 1.2 % for a 1.3 MeV γ ray. Gamma-ray spectrum with high-S/N was obtained by using the information of PPAC, which can detect RI beams, placed at 1.5 m upstream of the secondary target. Time difference between PPAC and Ge detectors [TOF(PPAC-Ge)], as shown in the inset of Fig. 3. Peak in this TOF(PPAC-Ge) spectrum can be assigned to be events that γ rays were emitted when the RI beam bombarded the secondary target. By gating on this peak, γ rays emitted from the fusion products can be clearly observed with high S/N. Figure 3 shows γ-ray singles spectra gated on time range of (a) all region and (b) peak in TOF(PPAC-Ge) spectrum. Gamma-ray peaks indicated by closed and open circles are associated with the transitions of ^{142}Pr and ^{141}Pr produced by the reactions of ^{130}Te(^{17}N,5n) and ^{130}Te(^{17}N,6n), respectively. By the selection of events correlating with RI beam, as shown in Fig. 3 (b), transitions of ^{142}Pr and ^{141}Pr can be clearly observed by the reduction of background components up to 20 %. In this work, with the event selection correlating with RI beam using the TOF(PPAC-Ge) information, analysis of prompt- and delayed-$\gamma\gamma$ coincidence data could be successfully performed with high-S/N and enabled improvement of level scheme of this odd-odd nucleus ^{142}Pr. All results are summarized in Ref. 3.

2.3. *Study for high-spin states in* ^{136}*Ba*

The experiment was carried out to study high-spin states in ^{136}Ba, one of N=80 isotones, using the ^{124}Sn(^{17}N, p4n) secondary fusion reaction.[4] The Ge-detector array was constructed using 12 Euroball tapered Ge detectors and 8 BGOACS's, as shown in Fig. 4. The total detection efficiency was obtained to be around 4 % for a 1.3 MeV γ ray. Si-ball consisting of 30 ΔE-type Si detectors with thickness of 170 μm was installed in the center of this Ge-detector array to measure the evaporated charged particles from the fusion products. Gamma-rays emitted from the isotopes of Ba and Ce, which were produced by the evaporation channels including proton and α particles, respectively, can be enhanced in the γ-ray spectrum by selecting events with Si-ball information. Analysis is now in progress.

Fig. 4. Ge-detector array consisting of Euroball tapered Ge detectors.

References

1. T. Shimoda *et al.*, Nucl. Instrum, Methods, B **70**, 320 (1992).
2. S. Mitsuoka *et al.*, Nucl. Instrum, Methods, A **372**, 489 (1996).
3. A. Takashima, master thesis, Osaka University (2010).
4. Y. Ito, master thesis, Osaka University (2011).

The super-allowed Fermi type charge exchange reaction for studies of isovector non-spin-flip monopole resonance

Y. Sasamoto*, T. Uesaka, S. Shimoura, S. Michimasa, S. Ota, H. Tokieda, H. Miya, S. Kawase, Y. Kikuchi, K. Kisamori, M. Takaki, M. Dozono, H. Mathubara

Center for Nuclear Study, Graduate School of Science, University of Tokyo, RIKEN campus, Wako, Saitama 351-0198, Japan
**E-mail: yoshiko@cns.s.u-tokyo.ac.jp*

K. Yako, S. Noji, K. Miki

Department of Physics, University of Tokyo, Hongo, Tokyo 113-0033, Japan

H. Sakai, T. Kubo, Y. Yanagisawa, H. Takeda, K. Yoshida, T. Ohnishi, N. Fukuda, D. Kameda, N. Inabe, N. Aoi, S. Takeuchi, T. Ichihara, H. Baba, S. Sakaguchi, P. Doornenbal, H. Wang, R. Chen, Y. Shimizu

RIKEN Nishina center, Wako, Saitama 351-0198, Japan

T. Kawahara

Department of Physics, Toho University, Funabashi, Chiba 274-8510, Japan

T. Kawabata, N. Yokota

Department of Physics, Kyoto University, Kyoto 606-8502, Japan

Y. Maeda, H. Miyasako

Fuculty of Engineering university of Miyazaki, Miyazaki, Miyazaki 889-2192, Japan

G. P. A. Berg

Department of Physics and Joint Institute for Nuclear Astrophysics, University of Notre Dame, Notre Dame, IN 4655, USA

In order to establish the HICE using the super-allowed Fermi type reaction that is selective to isovector the non-spin-flip mode, the ^{90}Zr(^{10}C,^{10}B(IAS)) experiment at 200 MeV/u was performed. We successfully observed the signature of the isovector non-spin-flip transition by detecting the gamma-ray emitted from ^{10}B(IAS). We also observed strength at around more than $Ex > 20$ MeV in ^{90}Nb coincident with 1022-keV gamma-ray which indicates the non-spin-flip transition.

Keywords: SHARAQ spectrometer, charge-exchange reaction (^{10}C,^{10}B), giant resonance

1. Introduction

Incompressibility of nuclear matter attracts a focus of experimental and theoretical studies because of its fundamental importance to define the equation of state of nuclear matter. Isoscalar incompressibility is studied through the isoscalar monopole resonances populated by the (α, α') reaction. To expand our knowledge to asymmetric nuclear matter, additional quantities are needed. Properties of isovector monopole resonances (IVMR) which are oscillation modes of the isovector density $\rho_{IV} = \rho_n - \rho_p$ are expected to be a key to understand asymmetric nuclear matter.[1]

In spite of its importance, however, experimental data on IVMR are scarce. Pion charge exchange reactions $(\pi^{\pm}, \pi^0)$[2] and heavy-ion charge exchange (HICE) such as (^{13}C, ^{13}N)[3,4] and (^{7}Li, ^{7}Be)[5] reactions have been used to study IVMR. These two-types of reactions provide inconsistent results of excitation energy, width, and/or multipolarity of the state.[3]

To understand nature of IVMR, a HICE probe selective to the non-spin-flip $(\Delta T = 1, \Delta S = 0)$ modes is essentially important. We propose the the super-allowed Fermi type charge exchange (^{10}C, ^{10}B(IAS)) reaction which is selective to $\Delta T = 1, \Delta S = 0$ modes. This selectivity is achieved only with the RI beam induced reaction as will be explained below.

We applied ^{90}Zr to the (^{10}C,^{10}B(IAS)) reaction at the ^{10}C beam energy of 200 MeV/u. In this report, we introduce the characteristics of the (^{10}C,^{10}B(IAS)) reaction and report the experimental result.

2. Characteristics of (^{10}C,^{10}B(IAS)) reaction

Our idea to establish a probe to isovector non-spin-flip states is based on use of a super-allowed Fermi transition between isobaric analog states in the projectile. This necessarily leads to a RI beam induced charge exchange reaction since light stable nuclei with $J^\pi = 0^+$ have no analog states because of their isospin $T = 0$. On the other hand, some unstable nuclei with $J^\pi = 0^+$ and $T = 1$ have isobaric analog states in neighbor nuclei. Gamma-rays emitted from IAS can be used to identify the super-allowed Fermi transition, experimentally.

The analog state of the ^{10}C ground state locates at $E_x = 1.740$ MeV in ^{10}B as shown in Fig.1. The transition from the ^{10}C ground state to the

1.740-MeV state in ^{10}B can be experimentally identified by observing the emitted γ-ray of 1022 keV. Thus, coincident detection of ^{10}B and 1022-keV γ-rays in the final state can be a clear signature of $0^+ \rightarrow 0^+$ transition in the projectile, and consequently, of the non-spin-flip excitation in the target.

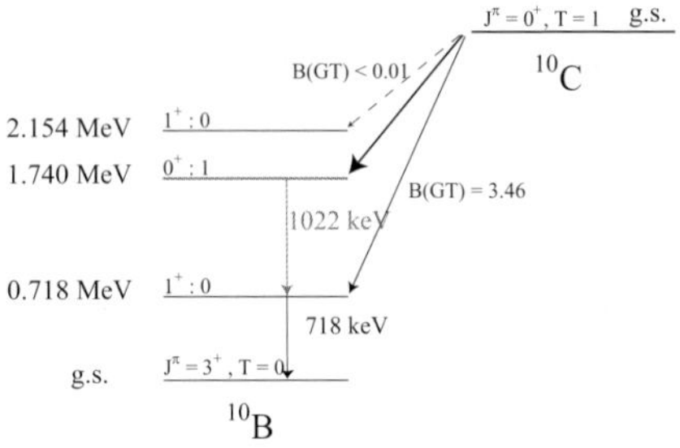

Fig. 1. Levels in ^{10}B. The analog state of the ^{10}C ground state is found at $E_x = 1.740$ MeV in ^{10}B.

3. Experiment and result

The experiment was performed at the RIBF facility at RIKEN using the SHARAQ spectrometer.[6] The secondary ^{10}C beam was produced as fragments of 250 MeV/u ^{14}N from a ^{9}Be target with a thickness of 20 mm positioned at BigRIPS-F0. A purity of approximately 95% was achieved for ^{10}C by using an Al wedge with thickness of 810 mg/cm^2. The intensity of the ^{10}C beam was approximately 2×10^6 pps at the secondary target. The beam trajectories were measured at the SHARAQ target to determine the position and angle of the beam. High-resolution beamline was set to the dispersion-matched mode so as to cancel the effect of energy spread of the beam itself.[7] A secondary ^{90}Zr target with a thickness of 150 mg/cm^2 was used. The emitted 1022-keV γ-rays to define the non-spin-flip transition were detected by using the NaI detector array (DALI2[8]) located near the secondary target.

Figure 2 shows the Doppler-corrected γ-ray spectrum, obtained by DALI2, coincident with ^{10}B particles in the SHARAQ focal plane. The 1022-keV peak can be clearly observed. Since this γ-ray is emitted from the 0^+ state at $E_x = 1.74$ MeV in ^{10}B, the non-spin-flip transition can be deduced by gating this peak. In Fig. 3, we can see the excitation energy in

^{90}Nb coincident with gamma-rays. The horizontal and vertical axes shows gamma-ray energy detected in DALI2 and excitation energy in ^{90}Nb respectively. At more than $Ex > 20$ MeV, strengths coincident with 1022-keV gamma-ray in ^{90}Nb are observed. These strengths are possible candidates of IVMR in ^{90}Nb.

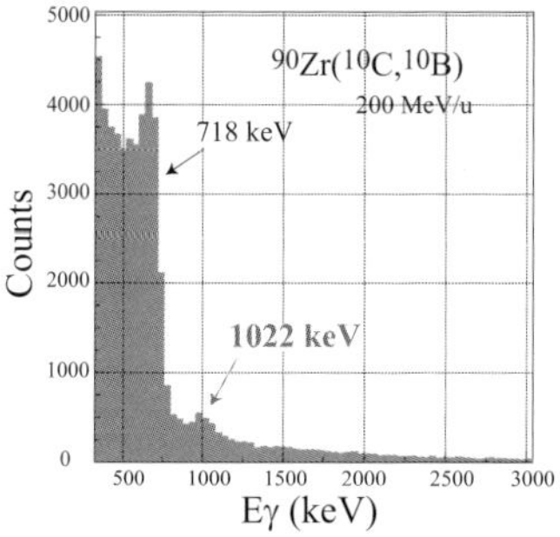

Fig. 2. Doppler-corrected γ-ray spectrum of ^{90}Zr(^{10}C,^{10}B) obtained by DALI2.

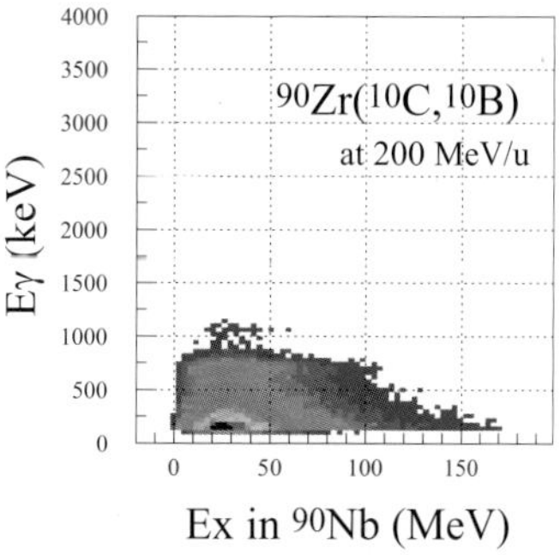

Fig. 3. Correlation of Doppler-corrected γ-ray energy and excitation energy in ^{90}Nb.

4. Summary

In order to establish the HICE using the super-allowed Fermi type reaction that is selective to the non-spin-flip mode, the ^{90}Zr(^{10}C,^{10}B(IAS)) experiment at 200 MeV/u was performed using the SHARAQ spectrometer at RIBF. We successfully observed the 1022-keV gamma-rays emitted from ^{10}B(IAS) which is a signature of the non-spin-flip transition. We also observed some strength around $Ex > 20$ MeV in ^{90}Nb coincident with 1022-keV gamma-ray, where the IVMR is expected to be exist.

References

1. G. Colo et. al., Phys. Rev. C **67**, 044306 (2003).
2. A. Erell et al., Phys. Rev. C **34**, 1822 (1986).
3. I. Lhenry, Nucl. Phys. A **599**, 245c (1996).
4. T. Ichihara et al., Phys. Rev. Lett. **89**, 142501 (2002).
5. S. Nakayama et al., Phys. Rev. Lett. **83**, 690 (1999).
6. T. Uesaka et al.: Nucl. Instrum. Meth. in Phys. Res. B **266**, 4218 (2008).
7. T. Kawabata et al.: Nucl. Instrum. and Meth. in Phys. Res. B **266**, 4201 (2008).
8. S. Takeuchi et. al.: RIKEN Accel. Prog. Rep. **36**, 148 (2003).

Invariant-mass spectroscopy of the unbound nucleus ^{13}Be

Y. Kondo, T. Nakamura, Y. Satou, Y. Hashimoto, N. Matsui, T. Nakabayashi, T. Okumura, M. Shinohara

Department of Physics, Tokyo Institute of Technology, Tokyo 152-8551, Japan

T. Matsumoto, N. Aoi, N. Fukuda, T. Gomi, M. Ishihara, H. Otsu, H. Sakurai, T. Sugimoto, S. Takeuchi, Y. Yanagisawa

RIKEN Nishina Center, Saitama 351-0198, Japan

N. Endo, M. Kitayama, T. Kobayashi, Y. Matsuda

Department of Physics, Tohoku University, Miyagi 980-8577, Japan

S. Kawai, Y. Togano

Department of Physics, Rikkyo Unibersity, Tokyo 171-8501, Japan

H. J. Ong, T. K. Onishi

Department of Physics, University of Tokyo, Tokyo 133-033, Japan

K. Ogata

Department of Physics, Kyushu University, Fukuoka 812-8581, Japan

S. Shimoura, M. Tamaki

Center for Nuclear Study (CNS), University of Tokyo, Saitama 351-0198, Japan

The unbound nucleus ^{13}Be was investigated by means of invariant-mass method in the one-neutron removal reaction of ^{14}Be at 69 MeV/nucleon. A p-wave resonance was observed at 0.51 MeV in the decay energy spectrum. The existence of the low-lying p state shows the quenching of the $N = 8$ shell gap.

Keywords: Unbound nucleus, invariant-mass method, intruder state

1. Introduction

Shell evolution is an important key word to understand structure of unstable nuclei located far from the stability since it causes anomalous phenomena such as disappearance of conventional magic numbers.[?,?,?] Shell evolution can be investigated from location of a non-normal parity state since it reflects the energy gap between major shells. For example, the ground state spin-parity of ^{11}Be is $1/2^+$ though the conventional shell structure gives $1/2^-$. This intruder $1/2^+$ ground state shows evidence of the quenching of the $N = 8$ shell gap between p and sd shells.

In this paper, we report observation of a low-lying non-normal parity state of the unbound nucleus ^{13}Be.[?] Unbound states of ^{13}Be were produced by the one-neutron removal reaction of ^{14}Be at 69 MeV/nucleon. Momentum distribution of ^{13}Be in the one-neutron removal reaction is useful to deduce the orbital angular momentum of a valence neutron.

2. Experiment and result

The experiment was performed at the RIKEN RI Beam Factory. A ^{14}Be secondary beam was produced by the projectile fragmentation of ^{18}O at 100 MeV/nucleon on a beryllium target and separated by the RIPS beam line.[?] The secondary ^{14}Be beam bombarded a liquid hydrogen target provided by a cryogenic proton and alpha target system (CRYPTA).[?] The incident angle of the beam was determined by two parallel plate avalanche counters (PPACs).[?] The outgoing charged particles and neutrons were separated by a dipole magnet. The charged particles were detected by two multi-wire drift chambers placed before and after the magnet, and plastic scintillator hodoscope. The outgoing neutrons were detected by a neutron detector array consists of 54 plastic scintillator bars. Details of the experimental setup can be seen in Ref. ?.

The invariant mass of the unbound nucleus ^{13}Be is reconstructed from the measured four-momenta of ^{12}Be and a neutron. Figure ?? shows the decay energy spectrum of ^{13}Be, which is well described by the fitting with three components (s, p, and d). The result of the fit is listed in Table ??. It is found that the 0.5-MeV peak in the decay energy spectrum corresponds to a p-wave resonance from the fitting of the decay energy spectrum and transverse momentum distribution of ^{13}Be. The width of the resonance at 0.5 MeV is consistent with the p-wave assignment. The spin and parity of the state is most probably $1/2^-$. A bump at around 2 MeV in the spectrum is assigned to a d-wave resonance. This assignment is consistent with the

Table 1. Result of the fit. a_s, E_r, Γ, and σ are scattering length, resonance energy, width, and cross section, respectively.

	a_s or E_r	Γ	σ
s	$-3.4(6)$ fm	—	21(3) mb
p	0.51(1) MeV	0.45(3) MeV	48(4) mb
d	2.39(5) MeV	2.4(2) MeV	34(3) mb

previous experiments.[?,?,?,?,?]

3. $N = 8$ shell gap

Figure **??** shows the energy difference $\Delta\epsilon$ between $1/2^+$ and $1/2^-$ states for $N = 7$ and $N = 9$ nuclei. Upper limit is shown for ^{13}Be since the peak position cannot be defined for a broad non-resonant s state. For the odd-odd nuclei ^{12}B and ^{14}B, the energy difference is obtained from the center of mass of the doublets $(2^-, 1^-)$ and $(1^+, 2^+)$ as in Ref. ?. For $N = 7$ nuclei, the energy difference between $1/2^+$ and $1/2^-$ is gradually decreases from the stable nucleus ^{13}C, and finally the order is inverted at the neutron-rich nucleus ^{11}Be. The same tendency can be seen for the $N = 9$ nuclei. The energy difference decreases from ^{15}C to ^{13}Be, showing the quenching of the $N = 8$ shell gap.

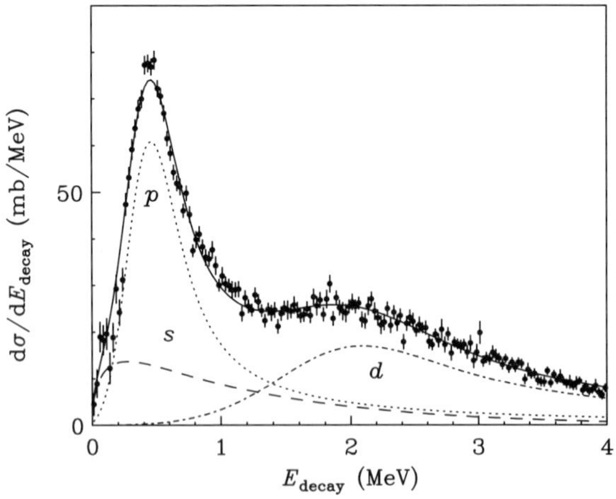

Fig. 1. Decay energy spectrum in the one-neutron removal reaction of ^{14}Be.

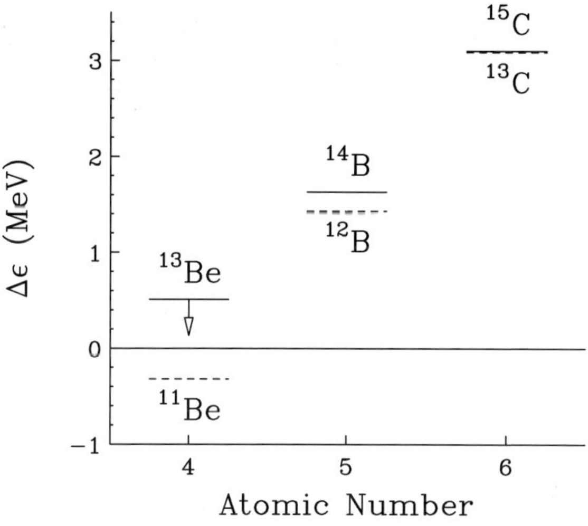

Fig. 2. Energy difference between $1/2^+$ and $1/2^-$ states for $N = 7$ and $N = 9$ nuclei shown by dashed and solid lines, respectively.

4. Summary

To summarize, a low-lying resonance of ^{13}Be was observed at 0.5 MeV above the neutron decay threshold. It is assigned to p wave from the analysis of the transverse momentum distribution of ^{13}Be and the width of the resonance. The existence of this intruder p state shows the quenching of the $N = 8$ shell gap.

GAMOW-TELLER TRANSITIONS IN PROTON RICH NUCLEI DEDUCED FROM THE COMBINED STUDY OF β-DECAY AND CHARGE-EXCHANGE REACTIONS

Y. FUJITA[a], B. RUBIO[b], B. BLANK[c], W. GELLETLY[d], T. ADACHI[a], J. AGRAMUNT[b], A. ALGORA[b], P. ASCHER[c], B. BILGIER[e], L. CÁCERES[f], R.B. CAKIRLI[e], E. GANIOĞLU[e], M. GERBAUX[c], J. GIOVINAZZO[c], S. GRÉVY[c], H. FUJITA[a], O. KAMALOU[f], H.C. KOZER[e], L. KUCUK[e], T. KURTUKIAN-NIETO[c], F. MOLINA[b], S.E.A ORRIGO[b], L. POPESCU[g], A.M. ROGERS[h], G. SUSOY[e], C. STODEL[f], T. SUZUKI[i], A. TAMII[i], J.-C. THOMAS[f]

[a] *Department of Physics, Osaka University, Toyonaka, Osaka 560-0043, Japan*
[b] *IFIC, CSIC-Universidad de Valencia, E-46071 Valencia, Spain*
[c] *Centre d'Etudes Nucléaires de Bordeaux Gradignan, Université Bordeaux 1, UMR 5797 CNRS/IN2P3, BP 120, F-33175 Gradignan, France*
[d] *Physics Department, University of Surrey, Guildford GU2 7XH, Surrey, UK*
[e] *Department of Physics, Istanbul University, Istanbul, Turkey*
[f] *Grand Accélérateur National d'Ions Lourds, BP 55027, F-14076 Caen, France*
[g] *SCK-CEN, Boeretang 200, B-2400 Mol, Belgium*
[h] *Physics Div., Argonne National Lab., 9700 South Cass Ave. Argonne, IL 60439, USA*
[i] *RCNP, Osaka University, Ibaraki, Osaka 567-0047, Japan*

Studying the weak nuclear response, especially the Gamow-Teller (GT) transitions, starting from stable as well as unstable pf-shell nuclei, is one of the key issues in nuclear and nuclear-astro physics. We study the GT transitions starting from $T_z = \pm 2$ mirror nuclei, respectively, by means of β decays and complementary hadronic $(^3\mathrm{He}, t)$ charge-exchange reactions. Under the assumption that isospin is a good quantum number, symmetry is expected for the structure of mirror nuclei and the GT transitions starting from them. The β-decay half-life and branching ratios and the strength distribution of GT transitions from the $(^3\mathrm{He}, t)$ reaction are compared and also combined for the understanding of the nuclear structure of pf-shell far-from-stability nuclei.

Keywords: Gamow-Teller transition, β decay, Charge-exchange reaction

1. Introduction

Studying the weak nuclear response of unstable pf-shell nuclei is one of the key issues in nuclear and nuclear-astro physics. The Gamow-Teller (GT)

transitions starting from stable as well as proton-rich unstable pf-shell nuclei play an important role in the core-collapse stage of type II supernovae.[?] However, our experimental study of the GT transitions starting from unstable nuclei with $T_z = -1$ or -2 is rather poor; accurate β-decay half-lives have been measured for several $T_z = -1$ and -2 nuclei,[?] but the determination of GT transition strengths, B(GT)s, have been out of the reach in decay studies. This is due to the small production rate of such far-from-stability nuclei, and also the small f-factors in the feedings to higher excited states, which makes the accurate determination of branching ratios difficult.

On the assumption that the nuclear interaction is charge-symmetric, isospin is a good quantum number. As a result, isobars with $\pm T_z$, where T_z defined by $(N - Z)/2$ is the z component of the isospin T, should exhibit a symmetric structure. Therefore, corresponding (analogous) GT transitions with $T_z = \pm 1 \rightarrow 0$ and $T_z = \pm 2 \rightarrow \pm 1$ are expected to have the same transition strengths. They can be studied by charge-exchange (CE) reactions and β decay.[?] It should be noted that the relative GT strengths can be well studied up to high excitation energy in CE reactions, but the absolute B(GT) strengths are obtained by normalizing to those obtained in β-decay studies. Thus, studies of β-decay and CE reaction are complementary.

A project to study the GT transitions starting from pf-shell nuclei far-from-stability is in progress. We have reported $T_z = +1 \rightarrow 0$ GT transitions on $A = 46, 50$ and 58 systems[?,?,?] and the studies of mirror $T_z = -1 \rightarrow 0$ transitions have also been partly reported.[?] Here we show the complementary $T_z = \pm 2 \rightarrow \pm 1$ studies for the $A = 56$ system. The study of GT strengths in the $T_z = +2 \rightarrow +1$ direction was performed by the ^{56}Fe(^{3}He, t)^{56}Co CE reaction at RCNP, Osaka, and the study in the $T_z = -2 \rightarrow -1$ direction was performed by the $^{56}\text{Zn} \rightarrow ^{56}\text{Cu}$ β decay at GANIL, Caen.

2. CE reaction at RCNP and β-decay study at GANIL

The ^{56}Fe(^{3}He, t)^{56}Co reaction was performed at the intermediate incoming energy of 140 MeV/nucleon and at 0°. This reaction is a unique tool for the study of GT transition strengths owing to the high selectivity of exciting GT states, high energy-resolution and also the close proportionality between the cross-sections at momentum transfer $q = 0$ and the B(GT) values[?]

$$\frac{d\sigma_{\rm GT}}{d\Omega}(0) = \hat{\sigma}_{\rm GT}(0) B({\rm GT}), \tag{1}$$

where $\hat{\sigma}_{\rm GT}(0)$ is a unit cross section for the GT transition at the momentum transfer $q = 0$ ($\approx 0°$). Although there are exceptions, the B(GT) values from the (^{3}He, t) CE reactions usually agreed within 5% with the β-decay B(GT) values by applying only one normalization parameter.[?]

The $T_z = +2 \to +1$, ^{56}Fe(^{3}He, t)^{56}Co spectrum[?] is shown in Fig. ??(a). The angular-distribution analysis suggests that most of the prominent levels are either $\Delta L = 0$, GT or Fermi states. An energy resolution of ≈ 20 keV has been realized by using a magnetic spectrometer in combination with the dispersion matching techniques.[?]

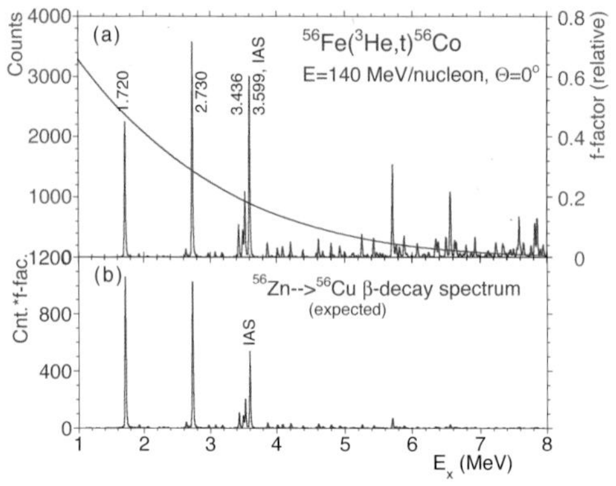

Fig. 1. (a) The ^{56}Fe(^{3}He, t)^{56}Co, $T_z = +2 \to +1$ spectrum taken at the incoming energy of 140 MeV/nucleon for events within scattering angles $\Theta \leq 0.5°$. Prominent states with $\Delta L = 0$, most probably the GT states, are indicated by their excitation energies in MeV. (b) The estimated ^{56}Zn β-decay spectrum obtained by multiplying the calculated f-factor with the ^{56}Fe(^{3}He, t) spectrum from figure (a).

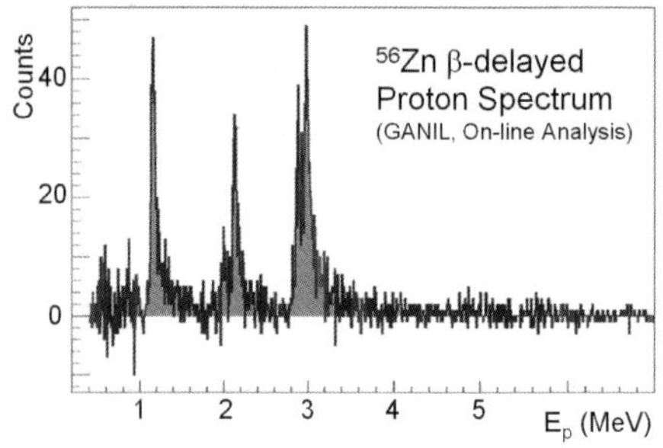

Fig. 2. The delayed-proton spectrum (on-line spectrum) obtained in the β-decay measurement of ^{56}Ni performed at GANIL. Note that the proton separation energy $S_p = 0.56$ MeV in ^{56}Cu. Therefore, the energy scale should be shifted by the amount of S_p to make a proper comparison with the ^{56}Co spectrum obtained in the ^{56}Fe(^{3}He, t) reaction.

The analogous $T_z = -2 \to -1$, GT transitions was studied by the ^{56}Zn$\to$^{56}Cu, β decay at the LISE3 facility[?] of GANIL. A ^{58}Ni^{29+} 74.5 MeV/nucleon beam with an average intensity of 3.7 eμA was accelerated by the two GANIL accelerators CSS1 and CSS2 and impinged on a ^{nat}Ni production target with a thickness of 265 μm placed at the entrance of the LISE spectrometer. The ^{56}Zn nuclei were produced using double-charge-exchange and the two neutron removal process, and were selected by the LISE spectrometer and the following Wien filter. The production rate was ≈ 2/min. The β-delayed proton and γ measurements were performed using a system of detectors including a DSSD and four EXOGAM Ge detectors.

The implantation detector setup at the focal plane consisted of a standard 300 μm-thick silicon detector providing the energy loss signal, a 300 μm double-sided silicon strip detector (DSSD) with 16 $\times$ 16 strips with a pitch of 3 mm used as the implantation device, and a 5 mm thick lithium-drifted silicon detector used as a veto detector for the implantation events and also to detect β particles. The implantation array was surrounded by four EXOGAM Ge detectors.

The $T_z = -2$ nucleus ^{56}Zn was first observed at GANIL.[?] In our measurement, the total number of implantations of ^{56}Zn was about 8,800, and we obtained an on-line value of $T_{1/2} = 33(2)$ ms for the decay of ^{56}Zn from the β-delayed proton measurement. This value is in agreement with the value of 30.0(17) ms obtained in the former β-delayed proton measurement performed at GANIL using the same LISE spectrometer.[?] The proton separation energy S_p of 0.56 MeV is low in the odd A nucleus ^{56}Cu. Therefore, the delayed-proton measurement was one of the key issues for the derivation of absolute B(GT) values. By reducing the thickness of the implantation DSSD detector down to 300 μm, we could reduce the fluctuation of the energy deposit of the emitted β particles. As a result, as we see in Fig. ??, a good energy resolution of $\approx$ 70 keV has been achieved.

3. Comparison: ^{56}Fe(^{3}He, t)^{56}Co reaction and ^{56}Zn β decay

Since the energy resolution ΔE of the delayed-proton measurement is approximately 70 keV and that of the (^{3}He, t) measurement is $\approx$ 20 keV, the comparison of the spectra from both measurements will be fruitful.

As mentioned, the peak intensities in the (^{3}He, t) spectrum is almost proportional to the GT transition strength B(GT), while the β-decay intensity is proportional to the value B(GT) multiplied by the phase-space factor f. The f-factors (relative values normalized to unity at $E_x = 0$) calculated following Ref.[?] as a function of decay Q value are shown by a smooth line in Fig. ??(a). Therefore, assuming good isospin symmetry of the $T_z = \pm 2 \rightarrow \pm 1$ GT transitions, the energy spectrum of ^{56}Zn β decay can be estimated by multiplying the ^{56}Fe(^{3}He, t) spectrum with the f-factor. The resultant spectrum is shown in Fig. ??(b).

By comparing the "estimated" β-decay spectrum with the ^{56}Zn β-delayed proton spectrum (Fig. ??), we see that the intensity ratios of the two low-lying GT transitions are similar. The proton peak at $E_p = 3.0$ MeV corresponds well with the complex of the IAS and a few GT states observed in the (^{3}He, t) measurement. However, we notice that the intensity is smaller than is expected.

It is known that the unit Fermi transition strength compared to the unit GT strength is larger by a factor of ≈ 5 in the β decay than in the $(^3\text{He}, t)$ reaction for mass $A = 56$ and at the incoming ^{3}He energy of 140 MeV/nucleon.[?,?] However, in reality, the IAS peak in the delayed-proton spectrum [Fig. ??] is not that big. It is suggested that the proton decay of the IAS is suppressed. If the decay of the $T = 2$, IAS in the $T_z = -1$ nucleus ^{56}Cu is examined from the view point of the isospin selection rule, we notice that the proton decay is energywise not allowed. The proton decay is possible only through $T = 1$ isospin impurities in the IAS. On the other hand, the proton decay of the $T = 1$, GT states are allowed. It is our interest whether we can see the γ decay from the IAS in the on-going off-line analysis.

4. Summary

As we have seen, the energies and the strengths of the $T_z = \pm 2 \rightarrow \pm 1$, mirror GT transitions in the $A = 56$ system were similar and consistent from a view point of isospin symmetry. This fact suggests that the nuclear structure of unstable proton-rich nuclei and the GT transitions starting from them can be deduced from the corresponding information that can be estimated by the $(^3\text{He}, t)$ measurement using the mirror nuclei as targets.

This work was in part supported by Monbukagakusho, Japan under Grant No. 18540270 and 22540310, Spanish MEC under Grant No. FPA2002-04181-Co4-03 and EURONS EU grant. YF and BR acknowledge the support of Japan-Spain collaboration program by JSPS and CSIC.

References

1. K. Langanke and G. Martínez-Pinedo Rev. Mod. Phys. **75**, 819 (2003).
2. C. Dossat et al., Nucl. Phys. **A792**, 18 (2007).
3. Y. Fujita, B. Rubio and W. Gelletly, Prog. Par. Nucl. Phys. **66**, 549 (2011), and references therein.
4. T. Adachi *et al.*, Phys. Rev. C **73**, 024311 (2006)
5. Y. Fujita *et al.*, Phys. Rev. Lett. **95**, 212501 (2005).
6. H. Fujita *et al.*, Phys. Rev. C **75**, 034310 (2007)
7. F. Molina, B. Rubio, Y. Fujita, and W. Gelletly, AIP Conf. Proc. **1265**, 49 (2010).
8. H. Fujita et al., OULNS annual report, Osaka University 2006, p. 91, unpublished.
9. R. Anne and A.C. Mueller, Nucl. Instrum. Meth. Phys. Res. B **70**, 276 (1992),
10. J. Giovinazzo et al., Eur. Phys. J. A **11**, 247 (2001).
11. D.H. Wilkinson and B.E.F. Macefield, Nucl. Phys. **A232**, 58 (1974).

SUPERNOVA NUCLEOSYNTHESIS, NEUTRINO MASS AND OSCILLATION, AND NUCLEAR WEAK INTERACTIONS

T. KAJINO[a,b], K. NAKAMURA[a], K. SATO[a,b], K. SHAKU[a,b], T. YOSHIDA[b], T. HAYAKAWA[c], S. CHIBA[c], D. YAMAZAKI[d], M.-K. CHEOUN[e], G. J. MATHEWS[f]

[a] *National Astronomical Observatory, 2-21-1 Osawa, Mitaka, Tokyo 181-8588, Japan*
[b] *Department of Astronomy, Graduate School of Science, University of Tokyo, 7-3-1 Hongo, Bunkyo-ku, Tokyo 113-0033 Japan*
[c] *Japan Atomic Energy Agency, Shirakawa-Shirane 2-4, Tokai, Ibaraki 319-1195, Japan*
[d] *Institute of Astronomy and Astrophysics, Academia Sinica, National Taiwan University, Roosevelt Road, Taipei 10617, Taiwan*
[e] *Department of Physics, Soongsil University, Seoul 156-743, Korea*
[f] *Center for Astrophysics, Department of Physics, University of Notre Dame, Notre Dame, IN 46556, U.S.A.*

We here discuss how to determine the total neutrino mass and oscillation parameters from supernova nucleosynthesis, the Galactic chemical evolution, and the CMB anisotropies. Neutrinos play the critical roles in nucleosynthesis of light-to-heavy mass nuclei in core-collapse supernovae. We study the nucleosynthesis induced by neutrino interactions and find suitable average neutrino temperatures in order to explain the observed solar system abundances of several isotopes ^{7}Li, ^{11}B, ^{138}La and ^{180}Ta. These isotopes are predominantly synthesized by the supernova ν-process. We also study the neutrino oscillation effects on their abundances and propose an astrophysical method to determine the unknown neutrino oscillation parameters θ_{13} and mass hierarchy.

Keywords: supernovae, nucleosynthesis, neutrino

1. Introduction

Still unknown mass and oscillation properties of neutrinos take the important keys to resolve many fundamental questions in physics and astronomy such as why baryon and lepton symmetries are broken in our Universe, why we need unified theory of elementary particles beyond the standard model, why the core-collapse supernovae explode, etc. Neutrino interactions with atomic nuclei play the critical roles in various scales of astronomical and cosmological phenomena. We discuss in this article how to determine the unknown neutrino oscillation parameters from the studies of element synthesis in supernovae and the Galactic chemical evolution.

2. Supernova Neutrinos

2.1. *Supernova Model*

Intensive flux of three flavor neutrinos with total energy $\sim 3.0\times10^{53}$ ergs are emitted from core-collapse supernovae.[1] This total energy is almost equal to the gravitational binding energy of a 1.4 $M_\odot$ neutron star. Our adopted presupernova model is a 16.2 $M_\odot$ star just before the supernova explosion which corresponds to SN 1987A.[2,3] Nucleosynthesis yields generally depend on supernova models with different zero-age main-sequence masses, metallicities, and explosion energies. Since the neutrino-induced nucleosynthesis is a very weak process, one can adopt a post-processing method in that we calculate only the explosive nucleosynthesis in well studied supernova model for SN1987A and then the contribution from the other supernova models scale to those calculated by Woosley and his collaborators.[4,5]

2.2. *Neutrino Spectrum*

Neutrinos are being emitted during $\sim$10 s after the supernova explosion. Both neutral and charged current interactions on abundant nuclei, ν-^{4}He and ν-^{12}C, spall them into free neutrons, protons, and other light nuclei. Subsequently the explosive nucleosynthesis called the ν-processes occurs at high-temperatures and densities when the shock arrives, which affects the abundances of the light elements ^{6,7}Li, ^{9}Be and ^{10,11}B.[6]

The $\nu_{\ ,\tau}$ and $\bar{\nu}_{\ ,\tau}$ bring about the production of ^{11}B from supernovae appropriate for the observed Galactic chemical evolution of the light elements.[7] The temperature of $\nu_{\ ,\tau}$ and $\bar{\nu}_{\ ,\tau}$ is thus estimated to be $T_{\nu_{\mu,\tau}} = 6.0 \pm 1.0$ MeV.[2] In our supernova model the neutrino energy spectra are assumed to obey Fermi-Dirac distributions with zero-chemical potentials. This temperature is also severely constrained from the measured isotopic ratio $(^{11}\mathrm{B}/^{10}\mathrm{B})_{\mathrm{sun}} = 4.5 \pm 0.1$ in the solar system abundance.

We use the nucleosynthesis of ^{180}Ta for the estimate of the temperatures of ν_e and $\bar{\nu}_e$. Both ^{138}La and ^{180}Ta are predominantly synthesized by the ν-process.[8] Although the calculated result can reproduce the solar abundance of ^{138}La with charged current reactions for ν_e temperature of 4 MeV, it overproduces the abundance of ^{180}Ta. Here we investigate the possibility that this overestimate originates from the unique feature that the naturally occurring abundance of ^{180}Ta is actually a meta-stable 9^- isomer with half-life of $\geq 10^{15}$ yr, while the true ground state is a 1^+ unstable state which β-decays with a half-life of only 8.15 hr (see Fig. 1).

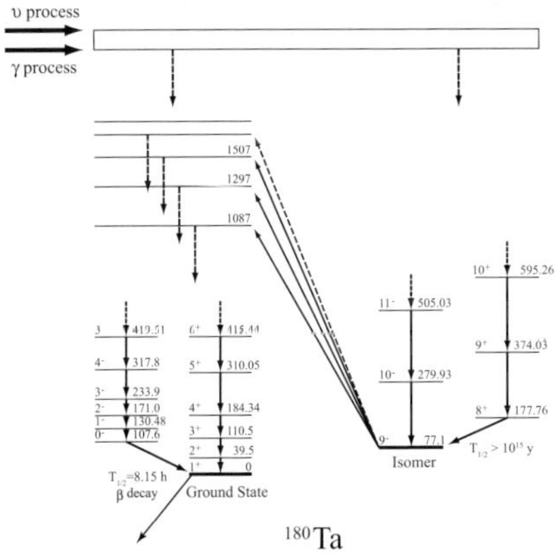

Fig. 1. Partial nuclear level scheme of ^{180}Ta.[10] The ground state decays via β-decay with a half-life of 8.15 h, while the isomer is a meta-stable state. The measured[9] excitation energies of states with weak transitions to the isomer are indicated.

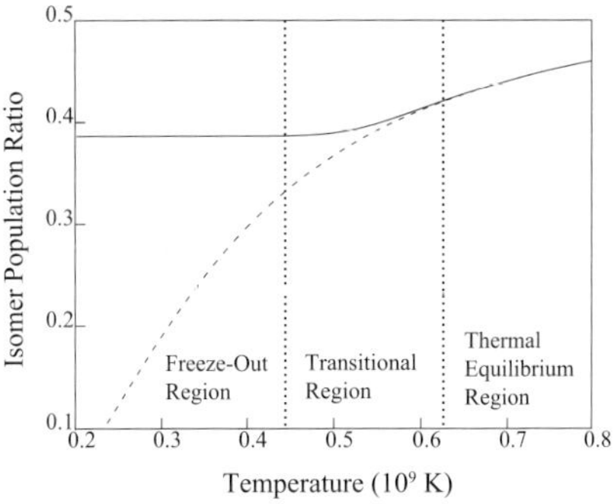

Fig. 2. Calculated isomer population ratio. The solid line denotes the time-dependent calculated result, and the dashed line denotes the result under the thermal equilibrium.

In the ν process, low-spin excited states in ^{180}Ta are strongly populated from ^{180}Hf by Gamow-Teller transitions and subsequently decay preferentially to the 1^+ ground state. However, in a high temperature photon bath, the meta-stable isomer is excited from the ground state by (γ,γ') reactions through highly excited states. Therefore, the final isomeric branching ratio should be evaluated by a time-dependent calculation, as displayed in Fig. 2. Our supernova nucleosynthesis calculation by including these complicated photon processes based on the measured nine linking transitions[9] shows that the isomeric residual population ratio turns out to be $\approx$ 0.39.[10] We finally obtain the concordant result between the observed solar system abundances of ^{138}La and ^{180}Ta and the theoretical predictions for $T_{\nu_e} \approx T_{\bar{\nu}_e} = 4.0$ MeV.[10]

Finally, we resolve a weak degeneracy between T_{ν_e} and $T_{\bar{\nu}_e}$ by imposing a neutron-rich condition $0.2 \leq Y_e < 0.5$ for successful r-process heavy element synthesis: $T_{\nu_e} = 3.2$ MeV and $T_{\bar{\nu}_e} = 4.0$ MeV.[11,12] This hierarchy applies also to the r-process nucleosynthesis as displayed in Fig. 3 that occurs in the gamma-ray bursts.[13]

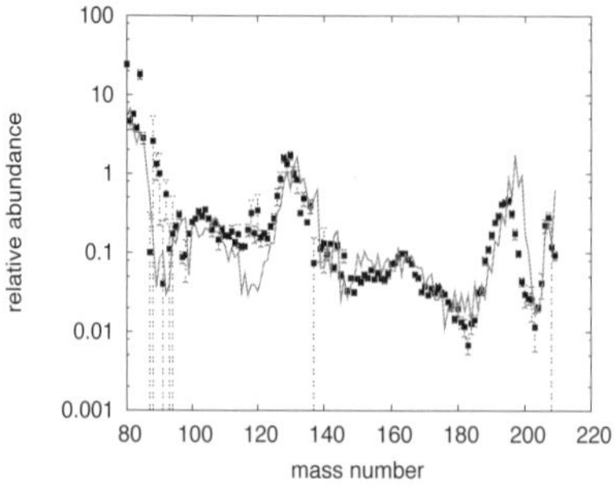

Fig. 3. Mass-averaged abundance (solid line) of the gamma-ray burst r-process[13] compared with solar system r-process abundance (points) of Käppeler et al. (1989). The mass-averaged abundance is normalized at ^{164}Er.

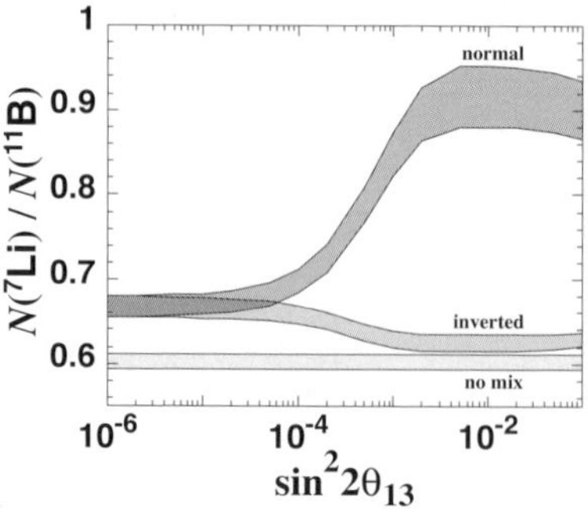

Fig. 4. The shaded ranges include the uncertainties of neutrino energy spectra deduced from the calculations using three sets of neutrino temperatures and total neutrino energies.

3. Neutrino Oscillation

3.1. *Oscillation Parameters*

Neutrino-flavor oscillations affect the nuclear reactions induced by charged current interactions, ν_e-A and $\bar{\nu}_e$-A, while neutral current interactions for all flavor of neutrinos do not change. We consider the change of the neutrino spectra due to three-flavor neutrino oscillations. The mixing parameters, except 13-mixing angle θ_{13} and the sign of $\Delta m^2_{31} = m^2_3 - m^2_1$, have been precisely determined from recent neutrino experiments:[14–16] $\Delta m^2_{21} = 7.9 \times 10^{-5}\mathrm{eV}^2$, $\Delta m^2_{31} = \pm 2.4 \times 10^{-3}\mathrm{eV}^2$, $\sin^2 2\theta_{12} = 0.816$, and $\sin^2 2\theta_{23} = 1.0$. The sign of Δm^2_{31} is still unknown, and we consider both of normal mass hierarchy ($m_1 < m_2 < m_3$) and inverted mass hierarchy ($m_3 < m_1 < m_2$). From CHOOZ experiment[17] upper limit of $\sin^2 2\theta_{13}$ has been determined: $0 \le \sin^2 2\theta_{13} \le 0.1$.

3.2. *Constraints from Nucleosynthesis*

In the explosive nucleosynthesis in supernovae, the reactions $^4\mathrm{He}(\nu, \nu'\mathrm{p})^3\mathrm{H}$ and $^4\mathrm{He}(\nu, \nu'\mathrm{n})^3\mathrm{He}$ are important for the production of ^{7}Li through $^3\mathrm{H}(\alpha, \gamma)^7\mathrm{Li}$ and $^3\mathrm{He}(\alpha, \gamma)^7\mathrm{Be}(e^-, \nu_e)^7\mathrm{Li}$ in He-leyer. If the ν-^{4}He reaction cross sections are enhanced due to the neutrino oscillations as to be discussed later, not only the abundance of ^{7}Li but also that of ^{11}B increase through the radiative alpha-capture reactions $^7\mathrm{Li}(\alpha, \gamma)^{11}\mathrm{B}$ and $^7\mathrm{Be}(\alpha, \gamma)^{11}\mathrm{C}(e^+\nu_e)^{11}\mathrm{B}$. The neutral current reactions, $^{12}\mathrm{C}(\nu, \nu'\mathrm{p})^{11}\mathrm{B}$ and

$^{12}C(\nu, \nu' n)^{11}C$, also are important for the production of ^{11}B. Details are discussed in refs. [2,18–20].

Figure 4 shows the predicted number-abundance ratio of $^7Li/^{11}B$ as a function of unknown mixing parameter $\sin^2 2\theta_{13}$ for both mass hierarchies. The uncertainty due to neutrino spectra is included as shaded regions. The $^7Li/^{11}B$ ratio in the case of adiabatic 13-mixing resonance and normal hierarchy is remarkably larger than that without neutrino oscillations, even with the spectral uncertainties included. Thus, the enhancement of observed $^7Li/^{11}B$ ratio may constrain the lowest value of θ_{13} and eliminate the possibility of inverted hierarchy.

Observational efforts[21,22] to obtain Li and B abundances would detect the signature of the ν-process in ^{11}B-enriched stars. Another observational information would be added from meteoritic studies of presolar X-grains produced from supernova ejecta.[23,24] The combination of SN nucleosynthesis theory and observations of light elements[25] may ultimately provide powerful constraints on mass hierarchy and the mixing angle θ_{13}.

4. Cosmological Constraint on Neutrino Mass

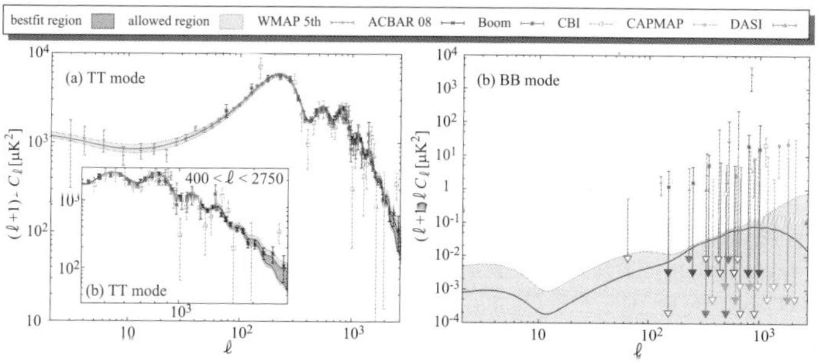

Fig. 5. Comparison of the best-fit computed total power spectrum with the observed CMB spectra. Plots show various ranges for: (a) TT($2 < \ell < 2750$), TT($400 < \ell < 2750$) and (b) BB($2 < \ell < 2750$) modes.[26]

We made a cosmological study of constraining the limits on the neutrino mass based upon the combined analysis of cosmic microwave background (CMB) anisotropies and the large scale structure in the presence of a primordial magnetic field (PMF)[26] . We find that a new upper bound on the neutrino mass turns out to be $\sum_{N_\nu} m_\nu < 0.8$ eV (from the 1σ constraint on the PMF) and < 2.2eV (from the 2σ constraint on the PMF) for $N_\nu = 3$. See ref. [26] for details.

This work has been supported in part by Grants-in-Aid for Scientific Research (20244035, 20105004) of the Ministry of Education, Culture, Sports, Science and Technology of Japan, and Heiwa Nakajima Foundation.

References

1. K. Hirata, et al. (Kamiokande Collaboration), *PRL* **58**, 1490–1493 (1987).
2. T. Yoshida, T. Kajino & D.H. Hartmann, *PRL* **94**, 231101 (2005); T. Yoshida, et al., *PRL* **96**, 091101 (2006); *Ap. J.* **649**, 319–331 (2006);
3. T. Shigeyama & K. Nomoto, *Ap. J.* **360**, 242–256 (1990).
4. R.D. Hofman, S.E. Woosley & A.A. Weaver, *Ap. J.* **549**, 1085–1092 (2001).
5. T. Rauscher, et al., *Ap. J.* **576**, 323–348 (2002).
6. S.E. Woosley, D.H. Hartmann, R.D. Hofman & W.C. Haxton, *Ap. J.* **356**, 272–301 (2000).
7. S.G. Ryan, et al., *Ap. J.* **549**, 55–71 (2001).
8. A. Heger, et al., *PL* **B606**, 258–264 (2005).
9. D. Belic, et al., *PR* **C65**, 035801 (2002).
10. T. Hayakawa, et. al., *PR* **C81**, 052801 (R) (2010); **C82**, 158801 (2010).
11. T. Yoshida, et al., *Ap. J.* **600**, 204–213 (2004).
12. K. Otsuki, H. Tagoshi, T. Kajino & S. Wanajo *Ap. J.* **533**, 424–439 (2000).
13. K. Nakamura, S. Sato, S. Harikae, T. Kajino & G.J. Mathews *Ap. J.*, submitted (2011).
14. Y. Ashie, et al. (Super-Kamiokande Collaboration), *PRL* **93**, 101801 (2004).
15. S.N. Ahmed, et al. (SNO Collaboration), *PRL* **92**, 181301 (2004).
16. T. Araki, et al. (KamLAND Collaboration), *PRL* **94**, 081801 (2005).
17. M. Apollonio, et al., *Eur. Phys. J.* **C27**, 331–374 (2003).
18. T. Yoshida, et al., *Ap. J.* **686**, 448–466 (2008); *Phys. Rev* **D80**, 125032 (2009); and references therein.
19. T. Suzuki, et al., *PR* **C74**, 034307 (2006); *PR* **C79**, 061603 (R) (2009).
20. M.K. Cheoun et al., *PR* **C81**, 028501 (2010); *PR* **C82**, 035504 (2010).
21. F. Primas, D. Dancan & J. Thorburn, *Ap. J.* **506**, L51–L55 (1998).
22. L. Rebull, et al., *The First Stars*, Springer, 2000, pp. 176.
23. P. Hoppe, et al., *Ap. J.* **551**, 478–485 (2001).
24. W. Fujiya, P. Hoppe & U. Ott, *Ap. J.* **730**, L7 (2011).
25. G.J. Mathews, T. Kajino, W. Aoki & W. Fujiya, arXiv:1108.0725v1, (2011).
26. D. Yamazaki, K. Ichiki, T. Kajino & G.J. Mathews, *PR* **D81**, 103519 (2010); *PR* **D81**, 023008 (2010).

KEK ISOTOPE SEPARATION SYSTEM FOR β DECAY SPECTROSCOPY OF R-PROCESS NUCLEI

Y.X. WATANABE, H. MIYATAKE, S.C. JEONG, H. ISHIYAMA, N. IMAI, Y. HIRAYAMA, M. OYAIZU

Institute of Particle and Nuclear Studies, High Energy Accelerator Research Organization (KEK), Tsukuba, Ibaraki 305-0801, Japan

K. NIKI, M. OKADA

Accelerator Laboratory, High Energy Accelerator Research Organization (KEK), Tsukuba, Ibaraki 305-0801, Japan

M. WADA, T. SONODA

Nishina Center for Accelerator-Based Science, RIKEN, Wako, Saitama 351-0198, Japan

Neutron-rich isotones with neutron shell closure N = 126 on the r-process path are supposed to be progenitors which form a peak around A = 195 in the r-abundance element distribution. Their β decay properties are considered to play critical roles for better understanding where the heavy elements such as gold and platinum were made in the universe. We are now advancing a project to construct KEK Isotope Separation System to study β decay properties and nuclear structures of neutron-rich nuclei around N = 126 experimentally. Those isotopes will be obtained via multi-nucleon transfer reactions induced by low-energy beams. For the study of the rare channel reaction products, it is essential to select both the element and mass number of them. We will adopt a gas catcher system combined with the laser resonance ionization and ISOL, which collect all reaction products and select the atomic number and mass of them efficiently. That system allows us to measure their β decay properties by two-layer plastic scintillators and germanium detectors.

1. Introduction

The rapid neutron capture process (r-process) in astrophysical environments, characterized by a high neutron density and a high temperature, is considered to be responsible for production of a large number of isotopes in the range of $70 \leq A \leq 209$ and also for synthesis of uranium and thorium in the universe [1]. The r-process network calculations usually include more than 4,000 nuclei in the region from the valley of β stability to the neutron drip line, where most of those nuclear physics properties are not known experimentally. As a good approximated analysis by assuming the (n, γ) - (γ, n) equilibrium and beta decay steady flow equilibrium (waiting point analysis) [2], however, the observed peaks of the solar r-abundance distribution can be characterized by the limited information of the waiting point nuclei with neutron shell closures on the r-process path.

The r-abundance peak around mass number 195 is considered to originate from the neutron shell closure with $N = 126$, and the ground state properties of those waiting point nuclei, such as beta decay half-lives, neutron separation energies and beta-delayed neutron emission probabilities, together with astrophysical simulations, are necessary to constrain temperature, neutron density, timescale, and neutrino flux at the r-process site. In the present r-process calculations, however, nuclear parameters for those waiting point nuclei fully rely on the theoretical predictions and the large deviations in those parameters from model to model are inevitable for such extremely neutron-rich nuclei. Therefore, experimental approach to the waiting point nuclei is an essential step in the search for the cosmic origins of the heavy elements.

We are now advancing a project to construct KEK Isotope Separation System at RIKEN for lifetime measurements of neutron-rich nuclei around neutron shell closure $N = 126$. In this project, those exotic nuclei will be accessed by using MNT (Multi-Nucleon Transfer) reactions, which have been pointed out to be most promising tool by several authors [3-5]. We will adopt the combination of ^{136}Xe beam and ^{198}Pt target as a good candidate for efficient production of $N = 126$ nuclei down to ^{200}W.

2. KEK Isotope Separation System

Figure 1 shows a schematic view of the KEK Isotope Separation System. The ^{136}Xe beam goes into a gas catcher system, where the ^{198}Pt target is located. Various nuclei produced by MNT reactions are stopped in a gas cell, and a certain element is selectively ionized by laser resonance ionization. Those ions are extracted by a high voltage, and are mass-separated by a high resolution bending magnet. They are transported to the detection system, where three detection stations with a tape transport system and detectors exist for lifetime measurements.

2.1. *Production of N=126 Nuclei*

The cross sections for production of nuclei around $N = 126$ by the MNT reactions of ^{136}Xe + ^{198}Pt was evaluated by the GRAZING calculation [6-8].

They show the maximum around the energies of 9 MeV/nucleon: ~100 μb for ^{202}Os and ~1 μb for ^{200}W. The expected yields were estimated by optimizing the target thickness, 12 mg/cm^2, for the beam energy of 9 MeV/nucleon and intensity of 2 pnA: ~5 pps for ^{202}Os and ~0.1 pps for ^{200}W.

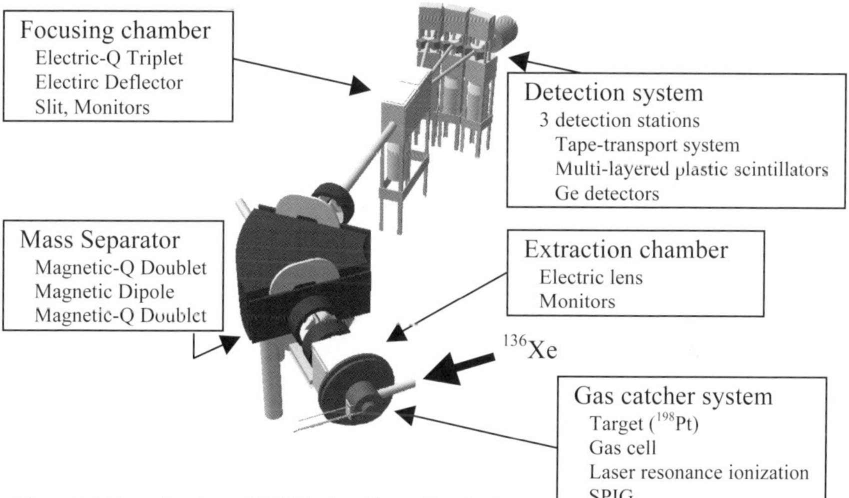

Figure 1. Schematic view of KEK Isotope Separation System.

2.2. *Gas Catcher System*

Since the various nuclei are generated by the MNT reactions, there is much contamination for extraction of a certain nuclide even by using a high resolution bending magnet. For example, there are 99.7% isobaric contaminations for extraction of ^{202}Os. Therefore, atomic number separation as well as mass number separation is essential for the lifetime measurements of such rare channel reaction products. Furthermore, since the ejected target-like fragments have a wide angular distribution of ~10 degrees around a large emission angle of 65 degrees, and have a wide energy spread from 0 to ~0.5 MeV/nucleon, it is difficult to separate and collect efficiently them by in-flight type electro magnetic spectrometer.

We have adopted a gas catcher system combined with laser resonance ionization technique and ISOL (Ion Separator On-Line) for effective collection and separation of the rare channel reaction products. Figure 2 shows a schematic view of the gas catcher system. The ^{136}Xe beam is injected into the gas cell, passing through the ^{198}Pt target window with thickness of 12 mg/cm^2. All reaction products are stopped and are almost neutralized in the gas cell filled with Ar gas of 500 mbar. The neutralized radioactive isotopes are rapidly transported to the outlet by the laminar gas flow, and they are re-ionized to singly-charged ions by laser resonance ionization technique [9]. Those ions are transported by the SPIG (SextuPole Ion Guide) [10], which is a cooling and focusing device, to the ISOL for mass separation. In this way, both of the atomic number (Z) and the mass number (A) separations are achieved efficiently by combination of the laser resonance ionization and ISOL.

After optimizing the gas cell design, the mean transport time and transport efficiency of the stopped atoms are estimated to be 253 ms and 56%, respectively, by the hydrodynamic calculations. Considering those values and assuming the laser resonance ionization efficiency of 17% [11], the total efficiency of the gas catcher system can be expected to be 5% for the ^{200}W nucleus, its half-life is estimated to be 423 ms by the KTUY mass formula [12].

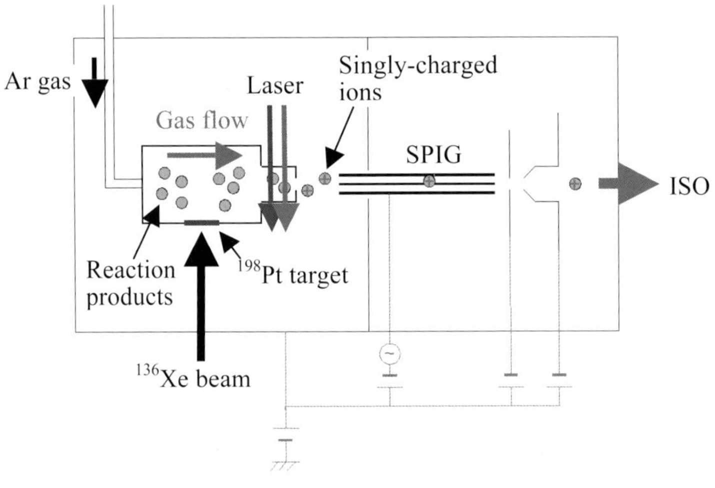

Figure 2. Schematic view of gas catcher system.

2.3. *Detection System*

The mass separated ions are transported to the detection system, which consists of three detection stations with a tape transport system and detectors for β-X coincidence measurements as shown in Fig. 1. Figure 3 shows a detail of the detection station. The exotic nuclei from the ISOL are implanted into a tape in a vacuum chamber surrounded by two-layer plastic scintillators placed at the atmosphere for β-ray detection (the efficiency, $\varepsilon \sim 80\%$). Three thin germanium detectors are used for the detection of X-rays ($\varepsilon \sim 60\%$ for 70 keV) and optionally γ-rays from the daughter nuclei are detected by other three thick

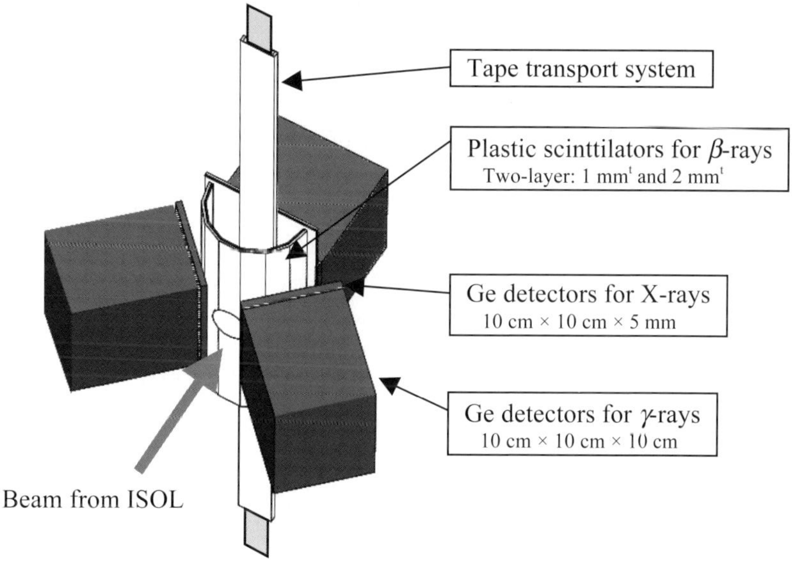

Figure 3. Schematic view of detection station.

germanium detectors ($\varepsilon \sim 20\%$ for 500 keV) for further nuclear structure investigation. Using three detection stations by switching the beam in rotation, the lifetimes of those exotic nuclei will be measured efficiently.

2.4. *Statistical Estimation*

In the case of ^{200}W, its production rate is $\sim 1 \times 10^4$ particles/day, 160 β-ray counts will be available for a one-day measurement, considering the collection efficiency (5%), the survival probability during the beam on time (40%) and the detection efficiency of β-rays (80%). The half-life will be determined with 10% error in those statistics.

2.5. *Research and Development*

The laser resonance ionization system has been constructed as a test bench in collaboration with RIKEN SLOWRI group, and new efficient ionization schemes of stable rhenium and iridium elements have been successfully searched by using two wavelength tunable dye lasers pumped by excimer lasers and the ionization chamber in which a filament of interest elements and a channeltron for ion detection are equipped. We found that the dye laser powers were enough to completely ionize those elements in the gas cell. The search for the ionization schemes of other stable elements of our interest with $Z = 68$ - 78 is in progress.

3. Summary

We will perform the lifetime measurements for neutron-rich nuclei around neutron shell closure $N = 126$, produced by MNT reactions of ^{136}Xe and ^{198}Pt.

The gas catcher system combined with laser resonance ionization and ISOL will be adopted to collect and separate the nuclei of interest. The rapid and efficient collection is enabled with the laminar gas flow, and the atomic number and the mass number are separated by laser resonance ionization and ISOL, respectively. After optimization of the gas cell design, the extraction efficiency is estimated to be 5% for the nucleus ^{200}W with expected half-life of 423 ms.

For the β decay lifetime measurements, tape transport system is employed. Using three detection stations suppresses the decay loss during beam-on time. The β decays of 160 counts will be detected for the nucleus ^{200}W and the lifetime will be determined with 10% error in a one-day experiment.

For research and development, we have found new efficient ionization schemes of stable rhenium and iridium elements by tuning the wavelength in the laser resonance ionization. The search for the ionization schemes of other stable elements of our interest with $Z = 68$ - 78 is in progress.

References

1. E.M. Burbidge *et al.*, *Rev. Mod. Phys.* **29**, 547 (1957).
2. K.-L. Kratz *et al.*, *Astrophys. J.* **403**, 216 (1993).
3. C.H. Dasso *et al.*, *Phys. Rev. Lett.* **73**, 1907 (1994).
4. V. Zagrebaev and W. Greiner, *Phys. Rev. Lett.* **101**, 122701 (2008).
5. L. Corradi *et al.*, *J. Phys. G: Nucl. Part. Phys.* **36**, 113101 (2009).
6. A. Winther, program GRAZING, http://personalpages.to.infn.it/~nanni/grazing (unpublished).
7. A. Winther, *Nucl. Phys. A* **572**, 191 (1994).
8. A. Winther, *Nucl. Phys. A* **594**, 203 (1995).
9. P. Van Duppen, *Nucl. Instrum. and Method B* **126**, 66 (1997).
10. H.J. Xu *et al.*, *Nucl. Instrum. and Method A* **333**, 274 (1993).
11. Yu. Kudryavtsev *et al.*, *Nucl. Instrum. and Method B* **204**, 336 (2003).
12. H. Koura *et al.*, *Prog. Theor. Phys.* **113**, 305 (2005).

SPIN MODES IN NUCLEI AND APPLICATIONS TO ASTROPHYSICAL PROCESSES

TOSHIO SUZUKI

Department of Physics, College of Humanities and Sciences, Nihon University
Sakurajosui 3-25-40, Setulaya-ku, Tokyo 156-8550, Japan
Center for Nuclear Study, University of Tokyo
Hirosawa, Wako-shi, Saitama 351-0198, Japan
**E-mail: suzuki@chs.nihon-u.ac.jp*

New shell model Hamiltonians, which give successful description of spin responses in nuclei, are applied to astrophysical processes in stars induced by nuclear weak interactions. Neutrino-nucleus reactions in supernova explosions, electron capture reactions in stellar core-collapse process and β-decays in r-process are discussed.

Keywords: Shell Model; Neutrino-Nucleus Reaction; Electron-Capture Reaction; β-decay; Supernova Explosion; R-Process.

1. New Shell-Model Hamiltonians in *p*-Shell and Neutrino-Nucleus Reactions

A new shell model Hamiltonian for *p-sd* shell, SFO,[1] has been obtainod from Cohen-Kurath and Millener-Kurath interactions by improving the spin-isospin dependent monopole component. The Hamiltonian can describe observed magnetic moments of *p*-shell nuclei systematically as well as Gamow-Teller (GT) transitions in ^{12}C and ^{14}C in configuration space including up to 2-3 $\hbar\omega$ excitations with the use of a small quenching for g_A; g_A^{eff}/g_A =0.95.[1]

The cross sections of the exclusive charge-exchange reaction, ^{12}C (ν_e, e^-) ^{12}N$_{g.s.}$, induced by the GT transition are well described by SFO, and they are found to be enhanced compared with those obtained by conventional Hamiltonians.[2] Neutrino-nucleus reaction cross sections on ^{12}C and ^{4}He induced by supernova neutrinos are also found to be enhanced, which leads to the enhancement of the production yields of light elements, ^{11}B and ^{7}Li, in supernova explosions. The reactions ^{12}C (ν, ν'p) ^{11}B and ^{7}Li (α, γ) ^{11}B are important to produce ^{11}B, while ^{7}Li is produced through the reaction chain ^{4}He (ν, ν'p) ^{3}H (α, γ) ^{7}Li. Enhanced cross sections are obtained for neutrino-induced reactions on ^{4}He with the use of more recent Hamiltonians such as WBP[3] compared with previous calculations.[4] The enhancement factor for the production yield is found to be about 20% and 30% for ^{11}B and ^{7}Li, respectively, for WBP+SFO Hamiltonians in a supernova explosion model (see Refs.[2] and[5] for the details). Heavy-flavor neutrinos with higher energies are important for these reactions, but charge-exchange reactions induced by ν_e become also important in case of matter oscillations.[5]

The SFO takes into account the important roles of the tensor components properly, that is, the sign rule for the orbit dependence of the monopoles is satisfied.[6] It is further improved in *p-sd* cross shell interactions, where the $\pi+\rho$ meson exchange interaction is taken for the tensor components.[7] The modified Hamiltonian, SFO-tls,[7] is applied to neutrino-induced reactions in ^{16}O. The reactions are induced dominantly by spin-dipole transitions. The spin-dipole transitions in ^{16}O for SFO-tls are found to reproduce well the energies of the spin-dipole states. The strength is more fragmented and shifted toward lower energies compared with that of SFO. The reaction cross sections for ^{16}O (ν_e, e^-) ^{16}F are enhanced for SFO-tls compared to SFO as well as CRPA[8] (about by 30$\sim$5% and 15% for SFO-tls and CRPA, respectively, at E_ν =30$\sim$60 MeV.)

2. Electron Capture Reactions in *fp*-Shell Nuclei

Precise evaluations of the electron capture rates in core-collapse proceeses at the end of the lives of the stars are quite important for the determination of the initial conditions of the supernova explosions and nucleosynthesis. New shell model Hamiltonians in *fp*-shell, GXPF1,[9] are used for the evaluation of electron capture rates in stellar environments. GT$_-$ strength in ^{58}Ni and M1 strengths in *fp*-shell nuclei are well reproduced for GXPF1J[10] with a universal quenching factor of g_A^{eff}/g_A =0.74 and g_S^{eff}/g_S =0.75. The $B(GT_+)$ for Fe and Ni isotopes are also reproduced rather well by GXPF1J.[11]

Electron capture rates at high densities, ρY_e =$10^7 \sim 10^{10}$ g/cm^3 with Y_e the lepton-to-baryon ratio, and at high temperatures, T_9 =1$\sim$10 with T =$T_9 \times 10^9$ K, are evaluated by using the GT transition strengths obtained by GXPF1J. The capture rates in ^{58}Ni and ^{60}Ni evaluated by using experimental GT strengths[12] are found to be well reproduced by the GT strengths of GXPF1J. They are improved compared to the rates obtained by conventional Hamiltonians, for example, KB3G[13] (see Ref.[11] for more details). Study of the GT transition strengths and electron capture rates in neutron-rich Ni isotopes as well as in other *fp*-shell nuclei such as Co and Mn isotopes are now under way.

3. R-Process Nucleosynthesis and β-Decays of N=126 Isotones

The β-decays of N =126 isotones are investigated by shell model calculations with the use of a modified G-matrix,[14] which describe well the energy levels of the isotones with Z =77∼81. Half-lives of the isotones with Z =64∼72 are evaluated by including both the GT and first-forbidden (FF) transitions.[15] Here, a common value for the quenching of g_A^{eff}/g_A =0.70 is used for both the transitions. The FF transitions reduce the half-lives obtained with the GT transitions only about by 1.5∼4 times for Z =62∼72, which are found to be short compared to those of Ref.[16]

The effects of the short half-lives of the β-decays on r-process nucleosynthesis are studied for constant entropy wind model. We found a robust slight shift of the third peak of the r-process element abundances toward higher mass region in various astrophysical conditions on supernova explosions[17] (detailed discussion is given in Ref.[17]).

4. Summary

New shell model Hamiltonians, which prove to give better description of spin modes in nuclei compared with previous calculations, are used to investigate nuclear weak transitions, and applied to astrophysical processes in stars.

(1) New neutrino-nucleus reaction cross sections for ^{12}C and ^{16}O are obtained by using SFO and SFO-tls Hamiltonian, respectively. The enhancement of the production rates of ^{7}Li and ^{11}B are found in light-element nucleosynthesis in supernova explosions. Charge-exchange reaction cross sections in ^{16}O are found to be enhanced compared with those of CRPA.

(2) Electron capture reaction rates in ^{58}Ni and ^{60}Ni are well described by a new shell model Hamiltonian, GXPF1J.

(3) Shell model calculations including both GT and FF transitions result in short half-lives for β-decays of N =126 isotones. This leads to a sligh shift of the third peak of the r-process element abundances toward higher mass region compared to the standard model.

We have thus shown some examples in which the progress of the studies in nuclear structure and reactions has important effects in astrophysical processes.

The author would like to thank M. Honma, T. Yoshida, S. Chiba, H. Mao, T. Kajino and T. Otsuka for the collaborations of the present work. This work has been supported in part by Grant-in-Aid for Scientific Research (C) 22540290 of the MEXT of Japan.

References

1. T. Suzuki, R. Fujimoto and T. Otsuka, *Phys. Rev.* C **67**, 044302 (2003).
2. T. Suzuki, S. Chiba, T. Yoshida, T. Kajino and T. Otsuka, *Phys. Rev.* C **74**, 034307 (2006).
3. E. K. Warburton and B. A. Brown, *Phys. Rev.* C **46**, 923 (1992).
4. S. E. Woosley, D. H. Hartmann, R. D. Hoffman and W. C. Haxton, *Astrophys. J.* **356**, 272 91990).
5. T. Yoshida, T. Suzuki, S. Chiba, T. Kajino, H. Yokomakura, K. Kimura, A. Takamura and D. H. Hartmann, *Astrophys. J.* **686**, 448 (2008).
6. T. Otsuka, T. Suzuki, R. Fujimoto, H. Grawe, and Y. Akaishi, *Phys. Rev. Lett.* **95**, 232502 (2005);
   T. Otsuka, T. Suzuki, M. Honma, Y. Utsuno, N. Tsunoda, K. Tsukiyama and M. Hjorth-Jensen, *Phys. Rev. Lett.* **104**, 012501 (2010).
7. T. Suzuki and T. Otsuka, *Phys. Rev.* C **78**, 061302(R) (2008).
8. E. Kolbe, K. Langanke and P. Vogel, *Phys. Rev.* D **66**, 013007 (2002).
9. M. Honma, T. Otsuka, B. A. Brown and T. Mizusaki, *Phys. Rev.* C **65**, 061301(R) (2002); **69**, 034335 (2004).
10. M. Honma, T. Otsuka, T. Mizusaki, M. Hjorth-Jensen and B. A. Brown, *J. Phys. Conf. Ser.* **20**, 7 (2005).
11. T. Suzuki, M. Honma, H. Mao, T. Otsuka and T. Kajino, *Phys. Rev.* C **83**, 044619 (2011); *Prog. in Part. Nucl. Phys.* **66**, 385 (2011).
12. M. Hagemann et al., *Phys. Lett.* **B579**, 251 (2004);
    N. Anantaraman et al., *Phys. Rev.* C **78**, 065803 (2008).
13. E. Caurier, G. Martinez-Pinedo, F. Nowacki, A. Poves and A. P. Zuker, *Rev. Mod. Phys.* **77**, 427 (2005).
14. S. J. Steer et al., *Phys. Rev.* C **78**, 061302 (2008);
    L. Ryndstrom et al., *Nucl. Phys.* **A512**, 217 (1990).
15. E. K. Warburton, J. A. Becker, B. A. Brown and D. J. Millener, *Annals of Physics* (NY) **187**, 471 (1988).
16. P. Moller, B. Pfeiffer and K.-L. Kratz, *Phys. Rev.* C **67**, 055802 (2003):
    P. Moller, J. R. Nix and K.-L. Kratz, *At. Data and Nucl. Data Tables* **66**, 131 (1997).
17. T. Suzuki, T. Yoshida, T. Kajino and T. Otsuka, submitted to *Phys. Rev.* C.

PROTON-CAPTURE NUCLEOSYNTHESIS IN NEUTRINO-DRIVEN SUPERNOVA EXPLOSIONS

SHINYA WANAJO[1,2], HANS-THOMAS JANKA[2], and SHIGERU KUBONO[3]

[1] *Technische Universität München, Excellence Cluster Universe, Boltzmannstr. 2, D-85748 Garching, Germany; shinya.wanajo@universe-cluster.de*

[2] *Max-Planck-Institut für Astrophysik, Karl-Schwarzschild-Str. 1, D-85748 Garching, Germany*

[3] *Center for Nuclear Study, University of Tokyo, RIKEN Campus, 2-1 Hirosawa, Wako, Saitama 351-0198, Japan*

We examine how the uncertainties in nuclear data inputs affect the νp-process in the neutrino-driven winds. We find that a small variation of the rates of triple-α and of the (n,p) reaction on ^{56}Ni leads to a substantial change in the p-nuclei production. The uncertainty in the nuclear mass of ^{82}Zr can also lead to a factor of two reduction in the abundance of the p-isotope ^{84}Sr.

1. Introduction

The astrophysical origin of the p-nuclei is not fully understood, in particular of light p-nuclei including ^{92,94}Mo and ^{96,98}Ru. The recent discovery of a new nucleosynthetic process, the νp-process, has improved this difficult situation.[1–3] In the neutrino-driven winds of core-collapse supernovae, $\bar{\nu}_e$ capture on free protons gives rise to free neutrons in the proton-rich matter. These neutrons induce the (n,p) reactions on the β^+-waiting point nuclei along the classical rp-process path (e.g., ^{64}Ge with the β^+-decay half-life of 1.06 min), which bypass these nuclei. It has been shown that the p-nuclei up to $A \sim 110$, including ^{92,94}Mo and ^{96,98}Ru, can be produced by the νp-process in neutrino-driven winds.[3]

There have been continuing experimental works relevant to the νp-process[4] since its discovery. However, we still rely upon theoretical or limited experimental estimates for the majority of nuclear reactions accompanied with the νp-process, which may suffer from uncertainties. There are also a number of isotopes without experimental mass measurements on the νp-process pathway.[5] In the following sections, we test the effect of uncertainties in some key nuclear ingredients (for more detail, see ref.[6]).

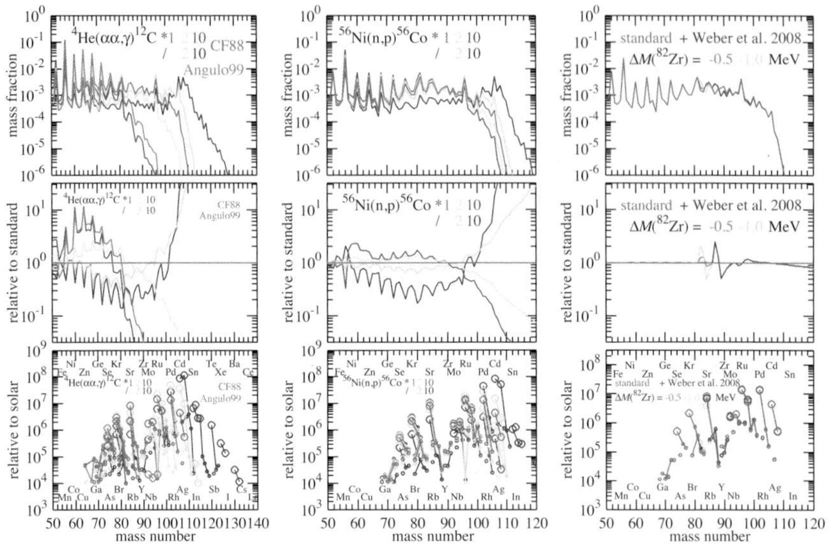

Fig. 1. Comparison of the nucleosynthetic results for various nuclear data inputs (left: triple-α rate; middle: $^{56}Ni(n,p)^{56}Co$ rate; right: nuclear masses). The neutrino-driven wind model is taken from ref.[6] The mass fractions (top) and their ratios relative to those for the standard model (middle) are shown as a function of atomic mass number. The bottom panels show the abundances of isotopes (connected by a line for a given element) relative to their solar values.

2. Triple-α rate

The triple-α process plays an important role for setting the proton-to-seed ratio Y_p/Y_h at the beginning of the νp-process. The result of sensitivity tests for the triple-α rate is shown in Fig. 1, where the forward and inverse rates are multiplied or divided by factors of 2 and 10. We find substantial changes in the production of p-nuclei with $A \sim 100 - 110$ for a factor of 2 variation on the rate, and more drastic changes for a factor of 10 variation. A larger triple-α rate leads to a more efficient seed production and thus smaller Y_p/Y_h. As a result, efficiency of the νp-process for heavy element synthesis decreases.

We find that a replacement of the triple-α rate (CF88)[7] by that of Anglo99[8] inhibits production of p-nuclei for $A > 80$ (Fig. 1). In fact, the net effect of including the 9.2 MeV state (not confirmed by recent experiments) in Angulo99 corresponds to the rate of CF88 multiplied by a factor of 10. This demonstrates the importance of future re-evaluations of (possible) contribution from higher levels than the 7.64 MeV state in ^{12}C.

3. (n, p) reaction on the seed nuclei

The νp-process starts from the seed nucleus ^{56}Ni, which is formed from free nucleons earlier. We find a remarkable change in the p-abundances with $A \sim 110$ by a factor of 10 with only a factor of 2 variation on ^{56}Ni$(n,p)^{56}$Co (Fig. 2). This demonstrates that the (n,p) reaction on the first (n,p)-*waiting nucleus* ^{56}Ni plays a key role for the progress of nuclear flows.

It should be noted that a smaller rate leads to more efficient νp-processing. The reason can be explained as follows. We find that the νp-process path proceeds through even-even $Z = N$ isotopes and deviates from ^{84}Mo ($Z = N = 42$) toward $Z < N$, reaching ^{96}Pd on the $N = 50$ shell closure. These isotopes have large abundances during the whole νp-process phase. The top panel of Fig. 2 shows that the abundances with $A = 60, 64, 68, 72, 76, 80$, and 84 are similar to that of $A = 96$. The (n,p) rates on ^{56}Ni and ^{96}Pd, neutron magic nuclei on $N = 28$ and 50, respectively, are a factor of 4–10 times smaller than the others. This indicates that the free neutrons created by neutrino capture are preferentially consumed by the even-Z nuclei with $30 \leq Z \leq 42$ (that act as *neutron poisons*), rather than by ^{96}Pd.

A reduction of the ^{56}Ni(n,p) rate (by a factor of 2 or 10) reduces the abundances of these neutron poisons by a similar factor ($A \sim 60 - 90$, Fig. 2; middle). Despite this, the abundance of ^{96}Pd does not decrease (even increases). This is due to the faster (n,p) rates for the $Z = N$ $(= 30 - 42)$ nuclei, causing the nuclear flows from ^{56}Ni to immediately reach ^{96}Pd and to stagnate there. As a result, a larger number of free neutrons becomes available for the ^{96}Pd(n,p) reaction.

4. Nuclear masses on the νp-process pathway

A number of nuclear masses along the νp-process pathway were recently obtained by precision measurements,[5] including those of ^{92}Ru and ^{93}Rh. Here, we present the nucleosynthetic result with inclusion of these new masses. We confirm the suppression of the flow through ^{87}Mo$(p,\gamma)^{88}$Tc.[5] This leads to a factor of three enhancement of ^{87}Sr and a factor of two reduction of ^{89}Y, which are however not p-isotopes. The other p-abundances, including of ^{92,94}Mo, are almost unchanged.

Our calculations of sensitivity tests for all other unmeasured masses on the νp-process path show that the mass of ^{82}Zr (or the proton separation energy of ^{83}Nb) on $N = 42$ plays an important role for production of a light p nuclei ^{84}Sr. We find that a 1.0 MeV reduction of the ^{82}Zr mass (equivalent to a reduction of the proton separation energy of ^{83}Nb) leads to a reduction of the ^{84}Sr abundance by a factor of two. An increase of the ^{82}Zr mass has no effect on the p-abundances.

This is particularly important when we consider the role of the νp-process to the solar inventory of most mysterious p-nuclei, ^{92}Mo and ^{94}Mo. As can be seen in Figs. 1–3 (bottom panels), the production factors of ^{92}Mo and ^{94}Mo are substantially smaller than the neighboring p-isotopes, in particular, ^{84}Sr. A reduction of ^{84}Sr would in part reduce this large gap.

5. Summary

We investigated the effects of uncertainties in nuclear data inputs on the νp-process in the neutrino-driven outflows of core-collapse supernovae.[6] Our result is summarized as follows; 1) Variation on the triple-α rate significantly affect the νp-process by changing $Y_{\rm p}/Y_{\rm h}$ at the onset of νp-processing. 2) Variation on the (n, p) reaction on ^{56}Ni (seed nuclei) shows great impact on efficiency of the νp-process for heavy element synthesis. 3) Application of the new experimental masses of ref.[5] exhibits a suppression of the flow ^{87}Mo$(p, \gamma)^{88}$Tc, which however do not affect the nucleosynthetic p-abundances. Our sensitivity tests for unmeasured nuclear masses indicate that a future measurement of the ^{82}Zr mass (and of ^{93}Nb with a large estimated error) could reduce the abundance of ^{84}Sr by a factor of two.

Acknowledgments

The project was supported by the Deutsche Forschungsgemeinschaft through Cluster of Excellence EXC 153 "Origin and Structure of the Universe" (http://www.universe-cluster.de).

References

1. Fröhlich, C., et al. 2006, *Astrophys. J.*, 637, 415
2. Pruet, J., Hoffman, R. D., Woosley, S. E., Buras, R., & Janka, H.-Th. 2006, *Astrophys. J.*, 644, 1028
3. Wanajo, S. 2006, *Astrophys. J.*, 647, 1323
4. Hayakawa, S., et al. 2010, 10th International Symposium on Origin of Matter and Evolution of Galaxies (OMEG2010), AIP conference proceedings, vol. 1269, 104
5. Weber, C., et al. 2008, *Phys. Rev. C*, 78, 054310
6. Wanajo, S., Janka, H.-Th., & Kubono, S. 2011, *Astrophys. J.*, 729:46
7. Caughlan, G. R. & Fowler, W. A. 1988, *At. Data Nucl. Data Tables*, 40, 283
8. Angulo, C., et al. 1999, *Nucl. Phys. A*, 656, 3

Momentum-dependent spin-current tensor contribution in collision dynamics

Yoritaka Iwata*

GSI Helmholtzzentrum für Schwerionenforschung,
Darmstadt, D-64291, Germany
**E-mail: y.iwata@gsi.de*

The momentum dependence of dissipation in heavy-ion reactions is studied. Special attention is paid to the dissipation due to the spin-orbit and spin-current tensor contributions in the Skyrme energy-density functional formalism. The spin polarization, which brings about large dissipation in the early stage of heavy-ion reactions, mainly arises from those two contributions.

Keywords: Spin-dependent force, dissipative dynamics

1. Introduction

The momentum-dependent dissipation in heavy-ion reactions is studied with the focus on the spin-polarization dynamics. Spin-polarization arises mostly from the spin-orbit force, and brings about large dissipation. It is important to note that dissipation due to the spin-orbit force should not be necessarily reduced to or similar to the wall-window type dissipation mechanism.[1,2]

The spin-current tensor and the spin-orbit contributions in the Skyrme energy density functional (EDF) are essential to the formation of the spin mean field, and give rise to the same dynamical effect, namely spin polarization. The spin-current tensor contribution, which acts like a modification of the spin-orbit contribution, consists of the contribution from the tensor force and the exchange part of the central force.[3] In this paper, attention is restricted to the former contribution of the spin-current tensor terms. Based on time-dependent density functional calculations, the momentum dependence of dissipation in heavy-ion reactions is studied for central collisions between two identical ^{16}O nuclei with respect to its correlation with spin polarization. For the mass dependence of spin polarization, see Iwata-Maruhn.[4]

2. Spin-polarization in collision dynamics

Dissipation due to the spin-polarization (see, Panel (a) of Fig. 1) is not necessarily similar to the wall-window type dissipation mechanism. The former is more similar to kinematic diffraction than to the scattering on the one-body potential wall; indeed, regardless of the existence of the one-body potential wall, the spin-polarization causes randomization, equilibration, and energy transfer from kinetic to internal excitation energies. Among other effects, motion perpendicular to the reaction axis, which largely contributes to the dissipation, is generated by these spin-dependent forces. Note that the dissipation due to the spin polarization is quite large and plays a decisive role in fusion events.[4]

2.1. *Spin polarization*

To evaluate the momentum-dependent dissipation, the amplitude of spin polarization is calculated for collisions with several incident energies. Measuring the spin polarization is equivalent to evaluating the effect from spin-dependent currents inside the composite nucleus. The amplitude of spin polarization is calculated by the superposition of spin for each single nucleon:

$$P(t,\mathbf{r}) = \rho(t,\mathbf{r})_{\uparrow} - \rho(t,\mathbf{r})_{\downarrow},$$

where $\rho(t,\mathbf{r})_{\uparrow}$ and $\rho(t,\mathbf{r})_{\downarrow}$ denote the densities with spin up and spin down, respectively.

2.2. *Spin-orbit coupling*

Concerning the spin polarization, it is reasonable to begin with the spin-orbit coupling. It is defined by the scalar triple product

$$\mathbf{L}\cdot\mathbf{S} = (\mathbf{r}\times\mathbf{p})\cdot\boldsymbol{S} = (\mathbf{p}\times\boldsymbol{S})\cdot\ \mathbf{r}, \tag{1}$$

where $\mathbf{L}$ and $\mathbf{S}$ denote angular momentum and spin angular momentum, respectively. In collision situations $\mathbf{r}\times\mathbf{p}$ is related to the impact parameter. Let the reaction plane be (x,z) with the initial collision direction z, and the direction perpendicular to the reaction plane be y (see Panel (a) of Fig. 1). For simplicity, the spin direction of the initial state is assumed to be parallel to the y-axis. In this setting, because only the z-component of $\mathbf{p}$ and the y-component of $\boldsymbol{\sigma}$ are non-zero, we have

$$\mathbf{L}\cdot\mathbf{S} = (p_y\ S_z - p_z\ S_y)\ x = -p_z\ S_y\ x. \tag{2}$$

We see that only the x-component of the vector $\mathbf{r}$ plays a role.

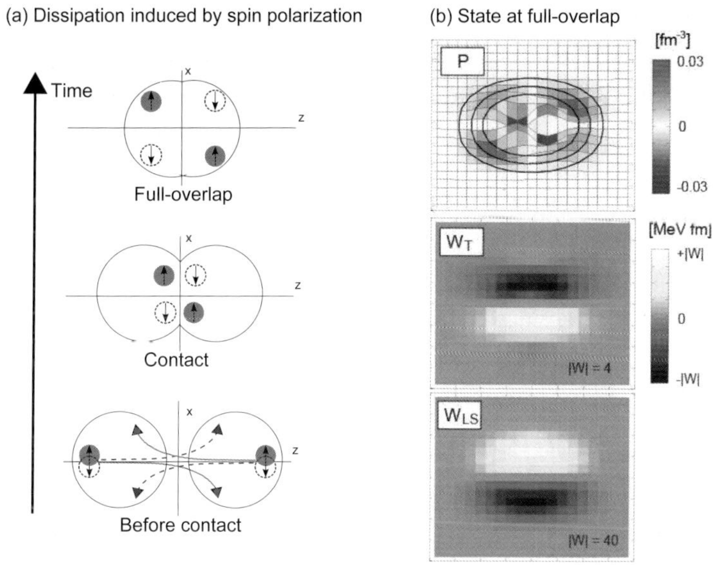

Fig. 1. Panel (a) illustrates the dissipation mechanism induced by the spin polarization, where the arrows denote the direction of the spins. The spin-dependent trajectories are shown in the bottom frame. In the full-overlap situation the spin is expected to be localized (top frame). Panel (b) shows the calculated full-overlap situation ($^{16}O + {}^{16}O$) for the case with the relative velocity of 20 % of the speed of light (the size of each frame is 16×16 fm^2). The spin polarization is shown in the topmost frame, where the density contours (each 0.05 fm^{-3}) are also depicted. The form factors of the spin mean-field are shown in the lower two frames.

2.3. *Spin mean-field*

The spin polarization mostly arises from the spin-orbit and spin-current tensor terms in the Skyrme-EDF.[5,6] Here we briefly review the mean-field formalism of these terms. Let ρ and $\mathbf{J}$ denote the number density and spin-current density, respectively. The contribution of the spin-current tensor and spin-orbit terms to the energy density functional has the form

$$\mathbf{W}_q(\mathbf{r}) \cdot (-i)(\nabla \times \boldsymbol{\sigma}) \tag{3}$$

where $\boldsymbol{\sigma}$ denotes the Pauli matrices, and $q = n, p$ (n and p stand for neutron and proton, respectively). $\mathbf{W}_q(\mathbf{r})$, which is called the form factor of the spin mean field, is decomposed into the spin-orbit and spin-current tensor contributions.

$$\mathbf{W}_q(\mathbf{r}) = \mathbf{W}_q^{LS}(\mathbf{r}) + \mathbf{W}_q^{T}(\mathbf{r}), \tag{4}$$

where $\mathbf{W}_q^{LS}(\mathbf{r})$ and $\mathbf{W}_q^{T}(\mathbf{r})$ denote the form factors of the spin-orbit and spin-current tensor mean fields, respectively. The contribution of the spin-orbit coupling is represented by

$$\mathbf{W}_q^{LS}(\mathbf{r}) = \tfrac{1}{2}W_0(\nabla\rho(\mathbf{r}) + \nabla\rho_q(\mathbf{r})),$$

where $\rho = \rho_p + \rho_n$, and W_0 is a constant. We deal only with the vector part of spin-current pseudo-tensor density. Those contribution is represented by

$$\mathbf{W}_q^{T}(\mathbf{r}) = \alpha\mathbf{J}_q(\mathbf{r}) + \beta\mathbf{J}_{q'}(\mathbf{r}) \tag{5}$$

with $q' = n, p$ satisfying $q \neq q'$, according to Stancu-Brink-Flocard.[3] In the following we refer to $\mathbf{W}_q^{T}(\mathbf{r})$ of Eq. (5) simply as "spin-current tensor terms", although they correspond, strictly speaking, only to a part of these terms.

Comparing Eqs. (1) and (3), $\mathbf{W}_q(\mathbf{r})$ in Eq. (3) plays the role of the vector $\mathbf{r}$ in Eq. (1), where the momentum $\mathbf{p}$ is replaced approximately by ∇ in the Skyrme-EDF. In this setting, the role of the spin-current tensor terms in the spin polarization can be evaluated by the corresponding x-component of $\mathbf{W}_q^{T}(\mathbf{r})$. In the following the x-components of $\mathbf{W}_q^{T}(\mathbf{r})$ and $\mathbf{W}_q^{LS}(\mathbf{r})$ are simply represented by $W_q^{T}(\mathbf{r})$ and $W_q^{LS}(\mathbf{r})$, if there is no ambiguity. The difference between spin-current tensor and spin-orbit contributions to the spin polarization is described by the ratio

$$W_q^{T}/W_q^{LS}(t) = \frac{\sum_{\mathbf{r}} |W_q^{T}(t,\mathbf{r})|}{\sum_{\mathbf{r}} |W_q^{LS}(t,\mathbf{r})|}, \tag{6}$$

which makes sense only if there is a certain similarity between the x-components of $W_q^{T}(\mathbf{r})$ and $W_q^{LS}(\mathbf{r})$ (otherwise attraction or repulsion happen irregularly from place to place, and this equation is not proper).

3. Momentum-dependent dissipation

Three-dimensional time-dependent density functional calculations are carried out in a spatial box $48 \times 48 \times 48$ fm^3 with a spatial grid spacing of 0.8 fm. The initial distance of the colliding nuclei is 20.0 fm; their initial positions were (0,0,10) and (0,0,-10). We consider central collisions between two identical ^{16}O nuclei with three different relative velocities: 10 %, 20 %, and 30 % of the speed of light. The SV-tls parameter set[7] is employed, which was lately introduced in the context of refitting the spin-current tensor contribution including only the exchange part of the central force.

The spin polarization is calculated for the full-overlap situations. According to the time evolution of the isoscalar dipole mode, the times

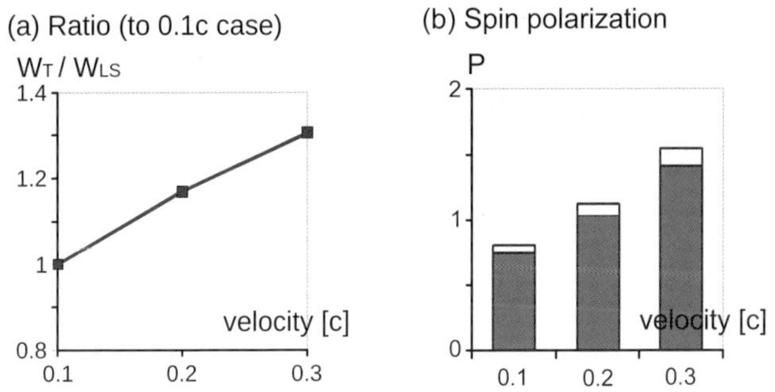

Fig. 2. Panel (a) shows the ratio between spin-orbit and spin-current tensor contributions depending on the relative velocity (values are normalized for the case with relative velocity of 10 % of the speed of light). Panel (b) presents the momentum-dependent spin polarization due to spin-orbit and spin-current tensor contributions, where the spin-orbit contribution corresponds to the total columns and the total spin polarization to the colored parts, so that the reduction of spin-polarization due to the spin-current tensor contribution can be seen in the uncolored parts of the columns. The proportionality between the spin polarization and the momentum is not valid in the following meaning; for the momentum-dependent spin polarization, it is difficult to draw a fitting line going through $(0, 0)$.

corresponding to the first full-overlap are estimated to be 9.5×10^{-22} s, 5.0×10^{-22} s and 3.5×10^{-22} s for the cases with relative velocity equal to 10 %, 20 %, and 30 % of the speed of light, respectively. The spin-current tensor and spin-orbit contributions are opposite in sign for all cases but similarly distributed in space (see Panel (b) of Fig. 1). Therefore the dissipation due to the spin-orbit contribution is reduced by the spin-current tensor contribution in this case (SV-tls).

Panel (a) of Fig. 2 shows the momentum dependence of the ratio between the spin-current tensor and the spin-orbit contributions in the full-overlap situation. It is found that the spin-current tensor contribution increases more rapidly compared to the spin-orbit contribution.

Panel (b) of Fig. 2 illustrates the expected dissipation (colored part of each column) and the corresponding components. It shows that the dissipation is positively correlated with the momentum, and almost 1 of 32 nucleons is fully spin polarized in every case. Despite the naive expectation (cf., Eq. (2)), the momentum-dependent dissipation of this kind is not exactly proportional to the momentum. The different dependence of the two contributions might be one reason for this effect, but the complicated nuclear medium effect should be taken into account in order to understand it.

4. Conclusion

The momentum-dependent dissipation has been studied for central collisions between two identical ^{16}O nuclei. It is suggested that the dissipation, which is calculated from the amplitude of spin-polarization, is positively correlated with the momentum, but not proportional to it. In particular the spin-current tensor contribution becomes larger for the collisions with higher momentum.

This work presents a new aspect of momentum-dependent dissipation, which is not necessarily similar to the wall-window type dissipation mechanism. Because the spin-orbit force is an important ingredient of the momentum-dependent nucleus-nucleus interaction, this work might contribute to the global understanding of the momentum-dependent nucleus-nucleus interaction.

The author thanks to Prof. J. A. Maruhn for fruitful discussions. This work was supported by the Helmholtz Alliance HA216/EMMI.

References

1. W. J. Swiatecki, Phys. Scripta **24** 113 (1981).
2. J. Randrup and W. J. Swiatecki, Nucl. Phys. **A429** 105 (1984).
3. Fl. Stancu, D. M. Brink, and H. Flocard, Phys. Lett. **B68** 108 (1977).
4. Y. Iwata and J. A. Maruhn, arXiv:1012.3363.
5. D. Vautherin, and D. M. Brink, Phys. Rev. **C5** 3 626 (1972).
6. Y. M. Engel *et al.*, Nucl. Phys. **A249** 215 (1975).
7. P. Klüpfel *et al.*, Phys. Rev. **C79** 034310 (2009).

STRONG NUCLEAR COUPLINGS AS A SOURCE OF COULOMB RAINBOW AND NEAR BARRIER FUSION SUPPRESSION

N. ALAMANOS†

CEA Saclay DSM/IRFU/DIR, F-91191 Gif-sur-Yvette Cedex, France

N. KEELEY†

Department of Nuclear Reactions, The Andrzej Sołtan Institute for Nuclear Studies, ul. Ho˙za 69, PL-00-681 Warsaw, Poland

1.1. *Introduction*

In July 2010 a paper was published by Alessia Di Pietro *et al.*, [1] on elastic scattering of ^{9,10}Be and quasi-elastic scattering of ^{11}Be from ^{64}Zn. The ^{11}Be results exhibit a non-Fresnel type pattern (Figure 1). Such a suppression of the Fresnel maximum was observed in the past in the case of elastic scattering on heavy targets and was attributed to strong Coulomb excitation effects. We were surprised by the results of Di Pietro *et al.* since ^{64}Zn is a relatively light nucleus for observing deviations from the standard Fresnel-type diffraction behavior due to Coulomb excitation effects. Indeed, Y. Kucuk *et al.*, [2] recently studied the elastic scattering of the halo nucleus ^{6}He from ^{58}Ni, ^{120}Sn, ^{144}Sm, ^{181}Ta and ^{208}Pb at incident energies close to the Coulomb barrier. They determined the range of the target charge (Z_T) where nuclear structure coupling effects become manifest. Their elastic scattering calculations display a marked deviation from the standard Fresnel-type diffraction behavior for targets with charge greater than $Z_T \approx 80$. This was attributed to strong Coulomb dipole breakup coupling effects

† Work partially supported by the LIA-COPIGAL .

generated by the Coulomb field of the heavy target. One may note that the lowest target charge for observing this effect, $Z_T \approx 80$, is much higher than the charge of the ^{64}Zn target (Z_T=30) used in Di Pietro *et al.*'s experiments.

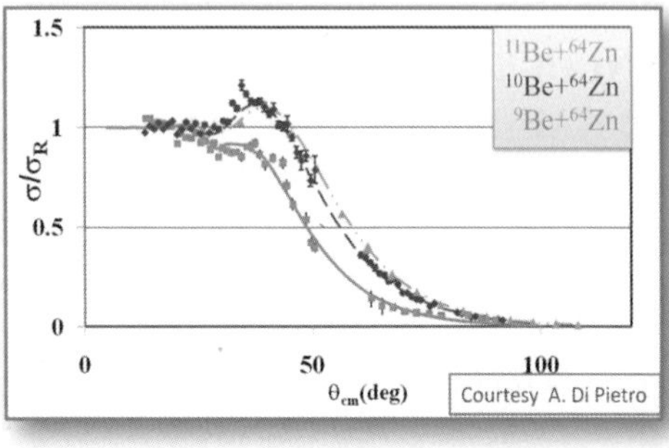

Figure1. (Color online). Elastic-scattering angular distributions on ^{64}Zn: ^{9}Be (triangles), ^{10}Be (diamonds), and ^{11}Be (squares). The lines represent the OM calculations for ^{9}Be (dot-dashed line), ^{10}Be (dashed line), and ^{11}Be (full line).

Taking into account the previous discussion, one can easily understand why the results of Di Pietro *et al*. came as a surprise to us. If our understanding was correct, the origin of the rainbow suppression in the ^{11}Be + ^{64}Zn elastic scattering could not be Coulomb coupling effects. The question that needs to be answered is then the following: What is the origin of the reduction of the Coulomb rainbow peak in the case of the ^{11}Be + ^{64}Zn data? The only possible scenarios we can foresee are strong nuclear coupling effects. If this is the case the observed behavior is exceptional. Until now we have never seen experimentally, at least to our knowledge, a disappearance of the Fresnel maximum due to purely nuclear couplings.

1.2. *On the polarization potential*

In the case of stable nuclei and in the absence of strong Coulomb excitation effects the overall success of folding model potentials for describing elastic-scattering data is well known [3]. This can be quantified by the value of the renormalization factor *N* of the real potential; *N* close to unity indicates that coupled-channel effects on elastic scattering are rather weak or concomitantly that the dynamic polarization potential (DPP) is rather weak. Indeed, coupled-channel effects can be simulated by an additional potential which is called the dynamic polarization potential. To a good approximation, this potential is real in the case of virtual couplings and imaginary in the case of real couplings [4]. In the case of weakly-bound nuclei the renormalization factor N is substantially decreased due to coupled-channel effects. For instance, in the case of the elastic scattering of ^{6,7}Li and ^{9}Be nuclei from various targets, the renormalization factor is N~0.6. This indicates that in the region of sensitivity of the optical potential, the dynamical polarization potential is of opposite sign and of a magnitude of the order of ~40% of the folding-model optical potential.

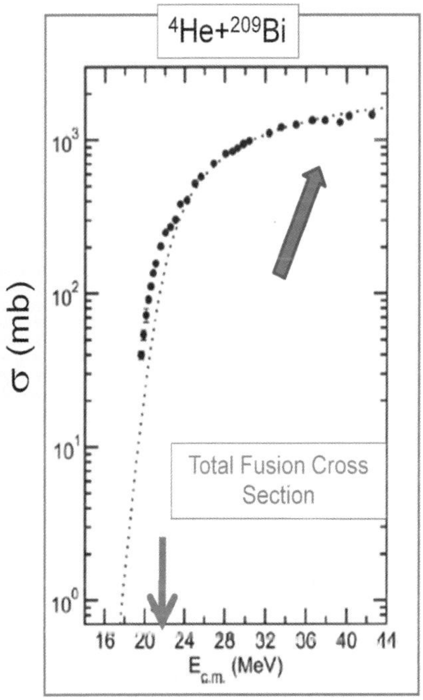

Figure 2. Calculated ^{4}He + ^{209}Bi fusion excitation functions compared with the total fusion data see. The arrows indicate the positions of the calculated Coulomb barriers.

To reproduce the experimental results in the absence of coupled-channel calculations the DPP has to be added to the theoretical folding-model optical potential. Many theoretical investigations, in particular for heavy-ion reactions, are under way, attempting to obtain a better understanding of the polarization potential.

1.3. *Near- and above-barrier fusion*

One interesting reaction channel for studying polarization potentials is near-barrier fusion. Formally, near-barrier fusion can be described by applying the incoming-wave boundary condition. In this framework the description of the sub-barrier fusion depends only on the real part of the nuclear potential. This is quite unique and allows specific coupled-channel effects or the real part of the polarization potential to be studied.

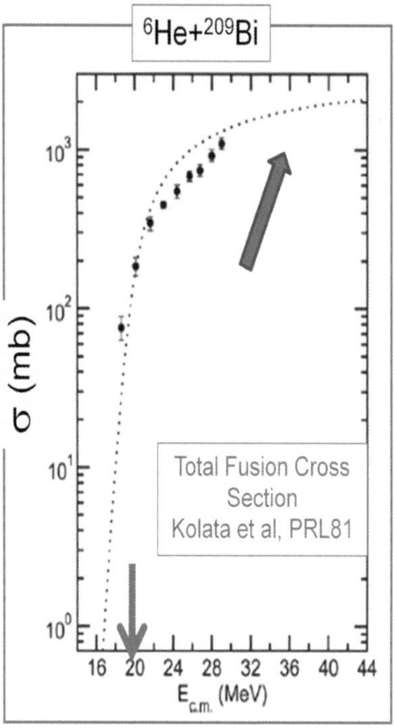

Figure 3. Calculated ^{6}He + ^{209}Bi fusion excitation functions compared with the total fusion data see. The arrows indicate the positions of the calculated Coulomb barriers.

We present in Figures 2 and 3 fusion excitation functions for the systems ^{4}He + ^{209}Bi and ^{6}He + ^{209}Bi calculated using double-folded potentials and compare them with data for total fusion. We would like to underline that in the present paper we will concentrate our attention on the above-barrier fusion. The data for the total fusion of ^{4}He + ^{209}Bi are rather well described for energies above the Coulomb barrier. The level of agreement obtained at above-barrier energies suggests that our adopted effective nucleon–nucleon interaction and target matter densities are physically reasonable. In contrast to ^{4}He + ^{209}Bi, the total fusion data for the halo nucleus system ^{6}He + ^{209}Bi are situated below the theoretical calculations. This is general behavior observed for the ensemble of total fusion data for halo nuclei systems [5]. Indeed, the most striking general conclusion with regard to total fusion of halo nuclei systems is not concerned with the sub-barrier regime but rather with the suppression of the measured total fusion cross-sections compared to the 'bare' no-coupling calculations at energies a few MeV above the nominal Coulomb barrier [5].

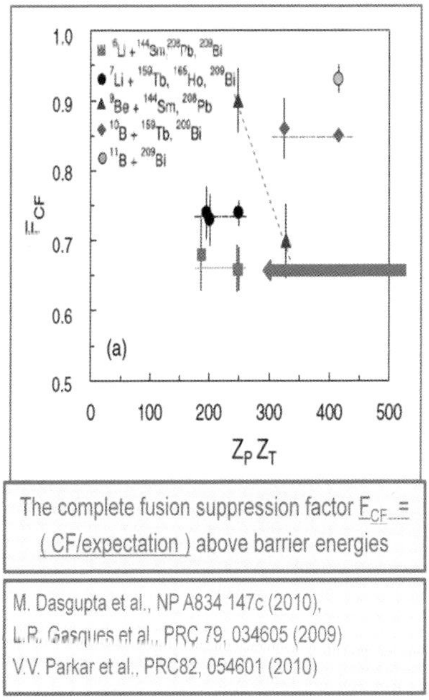

Figure 4. Complete fusion suppression factor at above barrier energies.

The study of the above-barrier suppression of complete fusion in reactions with weakly-bound stable or radioactive nuclei has recently attracted worldwide attention. M. Dasgupta *et al.* [6] have studied in a systematic way the above-barrier suppression of complete fusion in reactions with weakly-bound stable nuclei. In general the value of the suppression factor, defined as the ratio of the measured to the expected value of the above-barrier fusion, displays a strong correlation with the breakup threshold of the nuclei involved. This observation, in conjunction with new measurements of breakup at energies below the fusion barrier showing that breakup following transfer is the dominant mechanism, points once again to the importance of strong coupled-channel effects and concomitantly of the polarization potential.

We speculate that the study of total fusion at energies a few MeV above the nominal Coulomb barrier may be an alternative way to explore polarization potentials.

1.4. *Discussion of the ^{11}Be+ ^{64}Zn quasi-elastic scattering*

We have investigated the "exceptional" behavior of the quasi-elastic scattering of ^{11}Be+ ^{64}Zn by means of CDCC calculations employing a ^{10}Be + n cluster model of ^{11}Be [7]. In Figure 4 we compare the results of calculations including coupling to the ^{11}Be 1/2− first excited state only (2 channels), couplings to the full ^{10}Be + n continuum model space (24 channels), and the full breakup coupling scheme plus coupling to the ^{64}Zn (^{11}Be,^{10}Be)^{65}Zn single-neutron stripping reaction with the quasi-elastic scattering data of Di Pietro *et al.* [1].

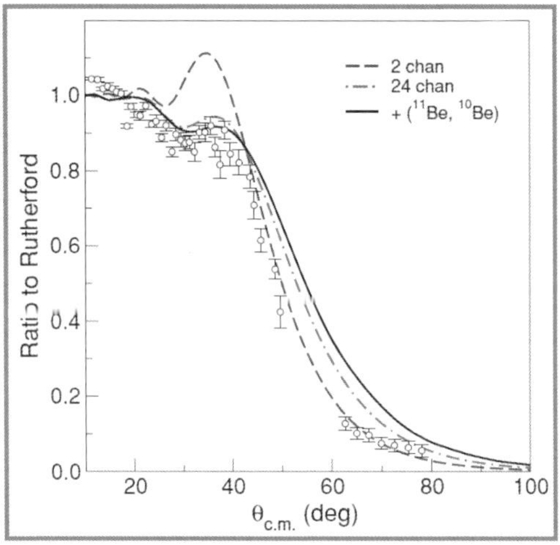

Figure 4. (Color online) Full model space CDCC calculations with both Coulomb and nuclear couplings, (solid curve), and with nuclear couplings only, but including the diagonal. Coulomb potentials (dot-dashed curve). The two-channel quasi-elastic scattering angular distribution is also included for reference (dashed curve). Note the linear cross-section scale.

Note that the cross section scale is linear rather than the usual logarithmic, to emphasize the region where the Coulomb rainbow would normally occur. The curves denote the calculated quasi-elastic scattering angular distributions, obtained from the sum of the elastic and inelastic scattering cross sections. The overall description of the data by the full CDCC calculation is reasonable, the agreement up to an angle of $\theta_{c.m.} \approx 45^\circ$ being excellent, with the calculation over-predicting the data at larger angles. Addition of the single neutron stripping coupling slightly improves the agreement at angles around $\theta_{c.m.} \approx 35^\circ$ but increases the cross section for $\theta_{c.m.} > 50^\circ$, leading to a somewhat worse description of the data for these angles. With these CDCC calculations we have shown that the most important source of the non-Fresnel-like shape of ^{11}Be+^{64}Zn quasi-elastic scattering is indeed nuclear coupling, as it is able to account for the majority of the effect by itself, although Coulomb coupling is still needed to obtain the best description of the data at angles around the Coulomb rainbow peak. We confirm therefore our initial expectation that this exceptional behavior of the ^{11}Be+ ^{64}Zn quasi-elastic scattering is essentially due to nuclear coupling effects.

In this paper we have discussed ^{11}Be+^{64}Zn quasi-elastic scattering and near-barrier fusion of halo and weakly bound nuclei. We have shown that in both cases coupled-channel effects play a very important role. We have speculated that the study of total fusion at energies a few MeV above the nominal Coulomb barriers may be an alternative way for exploring polarization potentials. Finally, we have shown that nuclear couplings are responsible for effacing the Fresnel-like shape in the case of ^{11}Be+^{64}Zn elastic scattering, which is a remarkable result.

References

1. A. Di Pietro *et al., Phys. Rev. Lett.* ***105****, 022701 (2010).*
2. Y. Kucuk *et al., Phys. Rev.* ***C79****, 067601 (2010).*
3. G. R. Satchler and W. G. Love, *Phys. Rep.* ***55***, 183 (1979).
4. G. R. Satchler, *Direct Nuclear Reactions (Clarendon Press, Oxford, 1983).*
5. N. Keeley *et al., Prog. Part. Phys.* ***63****, 396 (2009).*
6. M. Dasgupta *et al., Nucl. Phys.* ***A384*** *147C (2010).*
7. N. Keeley *et al., Phys. Rev.* ***C82****, 034606 (2010).*

ANALYTIC STUDY OF THE HINDERED FUSION
A LA RECHERCHE D'UNE FORMULE ANALYTIQUE POUR LA FUSION ENTRAVEE

Yasuhisa Abe
RCNP, Osaka University, Ibaraki, Osaka, 576-0047, Japan

Caiwan Shen
School of Science, Huzhou Teachers College, Huzhou, ZeJiang, 313000 China

David Boilley
GANIL, CEA/DSM-CNRS/IN2P3, Bvd Henri Becquerel, 14076 Caen, France and Université de Caen Basse-Normandie, F-14032 Caen Cedex, France

Bertrand B. Giraud
Institut de Physique Théorique, CEA/DSM, Centre de Saclay, Gif-sur-Yvette, 91190, France

Super-Heavy Elements (SHE) are synthesized by fusion reactions between heavy ions. Unfortunately, the fusion of very heavy systems is experimentally known to be strongly hindered, though the mechanism is not clearly understood yet. We have studied mainly the mechanism by numerical calculations, but recently have started the study with analytical method, which is supposed to clarify dynamical aspects of the physical mechanism. From where the hindrance comes, what the mechanism is, and to where the results are developed are briefly presented in the present talk.

1. Introduction

The fusion probability extracted from the measurements[1] clearly shows that the mass-symmetric entrance channel is hindered, while the mass-asymmetric one is almost normal in forming the same compound nucleus with Z= 102. In symmetric systems, the hindrance appears between ^{100}Mo-^{100}Mo and ^{110}Pd-^{110}Pd. However, there was no reaction theory which can explain or can predict the hindrance systematically. Present authors have proposed a mechanism and explained the experimental trends qualitatively with the Liquid Drop Model (LDM) energy surface and the Dissipation-Fluctuation or Diffusion dynamics.[2] Furthermore, the border between the normal and hindered fusions is clarified.[3] And with the fusion probabilities numerically calculated by Langevin equation, residue cross sections of SHE are predicted, in combination with the statistical theory of decay,[4] which is found to be in good agreement with the experiment recently performed.[5] An analytic solution with a simplified LDM potential is being obtained, which is expected to provide a transparent physical understanding and predictions of SHE production, as briefly recapitulated below.

2. Analytic Study of the Hindrance : One Dimensional Inverted Parabola

In fusion of heavy ions, we have to describe a dynamical process from the di-nucleus configuration made by the ions of the entrance channel and the spherical configuration of the compound nucleus. According to the LDM, the potential has a saddle point or a ridge line between the two configurations, irrespectively of parameterizations of those shapes. Fig. 1 shows its schematic view with respect to the distance (R) and the mass-asymmetry (α) of the entrance channel with the Two-Center-Parameterization. Since it is for very heavy systems, the saddle point is close to the spherical compound nucleus, which is natural with a fissility close to 1.0. It is worth to mention here that a small saddle point height does not mean easiness of overcoming, because nuclear collective motions are strongly dissipative. In other words, even if we increase incident energy more than the height, we cannot get the system fused. The incident kinetic energy is quickly dissipated into internal excitations, and is not left for mechanical motion. The situation is clearly demonstrated in 1-dimensional model in the following.[2]

One dimensional Langevin equation for the inverted parabola potential is given as

$$\mu\ddot{q} - \mu\omega^2 q - \gamma\dot{q} + R(t), \tag{1}$$

where the constant inertia mass μ and the constant friction coefficient γ are assumed for the relative distance q between two mass centers. Random force is denoted by $R(t)$, which is assumed to be Gaussian without memory (Markovian). Since the equation is linear, it can be solved analytically to give the probability to cross the potential barrier for any time t after the formation of the di-nucleus configuration located to a certain point of the

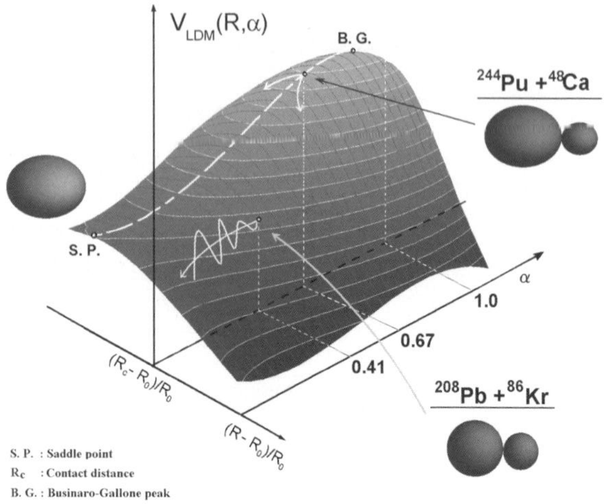

Fig. 1. Schematic LDM energy surface as function of the distance and the mass-asymmetry. Symmetric entrance configuration has to overcome the high ridge-line

slope,

$$P(t) = \frac{1}{2}\text{erfc}\left[-\frac{\langle q(t)\rangle}{\sqrt{2}\sigma_q(t)}\right], \tag{2}$$

where $\langle q(t)\rangle$ denotes an average trajectory and $\sigma_q(t)$ a variance of the distribution of the trajectories.

In the long time limit, it becomes a simple expression,

$$P(t\to\infty) = \frac{1}{2}\text{erfc}\left[\sqrt{\frac{x+\sqrt{x^2+1}}{2x}}\sqrt{\frac{B}{T}} - \frac{1}{\sqrt{2x(x+\sqrt{x^2+1})}}\sqrt{\frac{K}{T}}\right], \tag{3}$$

where B and K denote the potential height $\mu\omega^2 q_0^2/2$ and the initial kinetic energy $p_0^2/(2\mu)$, with q_0 and p_0 being the initial position and the the momentum. T denotes the temperature of the composite system, while x does $\beta/(2\omega)$ with β being γ/μ. It is readily understood that for the probability of the fusion to be 1/2, the initial kinetic energy is to be equal to

$$K_0 = (x+\sqrt{x^2+1})^2 \cdot B, \tag{4}$$

which figures out so-called extra-push energy,[6] because x is usually about a few to several. Note that K is a kinetic energy left for the di-nucleus configuration formed after overcoming of the Coulomb potential, which means that K is much smaller than the experimental incident energy. Overcoming process of the Coulomb barrier is also taken into account, which crucially affects the remaining K at the contact di-nucleus configuration.[7]

3. Border between the Normal and the Hindered Fusions

We know that nuclear systems are strongly dissipative, and thus expect that the kinetic energy K is negligible. Then, Eq. (3) tells us that the hindrance starts with non-zero B, i.e., that the border between hindered and non-hindered systems are given by the condition of B being equal to 0.

Fig. 2 left shows the border line for the symmetric systems in $Z-A$ plane of the incident ions.[3] The border is seen between ^{100}Mo and ^{110}Pd, consistent with the experimental observations. For asymmetric incident channels, the same analysis has been made.[8] It is worth to notice here that the results are obtained with the di-nucleus configurations of the full neck radius, i.e., with super deformed compound nuclear shape. That is based on the conclusion of our recent analyses that the time scale of the de-necking is much shorter than those of the relative distance and the mass-asymmetry.[9]

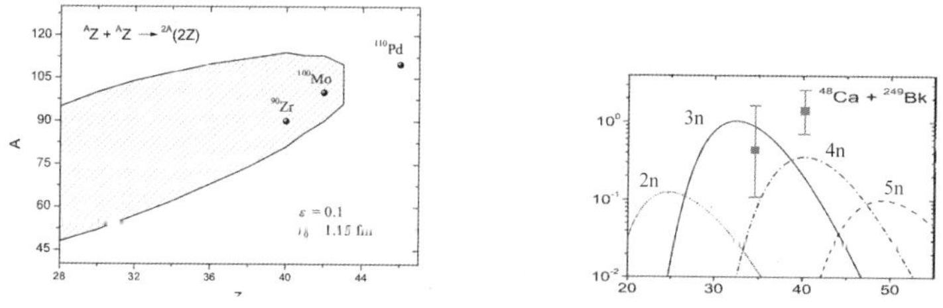

Fig. 2. Left: Borderline between hindered and non-hindered fusion (solid curve). The enclosed area is for the normal fusion, while the region outside is for the hindered fusion. Right: Predictions of residue cross sections are compared with the recent measurements

4. Predictions of the Excitation Functions for SHE production

An analytic formula for realistic multidimensional cases is under way. Thus, for the moment, we calculate the fusion probabilities numerically with Langevin equation. Combined with the statistical theory of compound nucleus decay,[4] we predict residue cross sections for SHE. An example of the prediction is shown in Fig. 2 right for ^{48}Ca + ^{249}Bk with the experimental data measured very recently at Dubna.[5] For $Z = 120$ compound nucleus, predictions are made with different incident channels.[10]

5. Summary Résumé

The hindrance of the fusion, la Fusion Entravée, d'où vient-elle ? Qu'est-elle ? Où va-t-elle ?
Answers are

(1) It stems from the strong dissipation of collective energy and the location of the di-nucleus configuration outside the ridge-line of the potential .
(2) It is not due to a mechanical motion, but due to a diffusion process.
(3) Analytic study provides an intuitive physical understanding as well as direct comparisons of cross sections of SHEs among various incident channels. The formula is under way.[11]

Acknowledgments

This work was supported in part by the LIA between GANIL and RIKEN - RCNP.

References

1. K.-H. Schmidt and W. Morawek, Rep. Prog. Phys. 54, 949 (1991) ; C.-C. Sahm et al., Nucl. Phys. A441, 316 (1985).
2. Y. Abe, D. Boilley, B.G. Giraud and T. Wada, Phys. Rev. E61, 1125 (2000).
3. C. Shen, Y. Abe, Q. Li and D. Boilley, Sciences in China, G52, 1458 (2009).
4. B. Bouriquet, G. Kosenko and Y. Abe, Eur. Phys. J. A22, 9 (2004) ; C.W. Shen, Y. Abe, D. Boilley, G. Kosenko and E.G. Zhao, Int. J. Mod. Phys. E17(suppl), 66 (2008); KEWPIE II, A. Marchix, PhD thesis, Univ. Caen, (2007), and HIVAP.
5. Yu. Ts. Oganessian et al., Phys. Rev. Lett. 104, 142502 (2010).
6. S. Bjornholm and W.J. Swiatecki, Nucl. Phys. A391, 471 (1982).
7. Y. Abe et al., Prog. Theor. Phys. Suppl. No. 146, 104-109 (2002).
8. C. Shen, D. Boilley, Q. Li, J. Shen and Y. Abe, Phys. Rev. C83 054620 (2011).
9. Y. Abe et al., Int. J. Mod. Phys. E17, 2214 (2008), and E18, 2169 (2009).
10. C. Shen et al., under preparation.
11. D. Boilley, Y. Abe, C. Shen and B. Yılmaz, CERN-Proc.-2010-001, 479 (2010); D. Boilley et al., publication under preparation.

STUDIES OF FISSION-YIELD MODELS

P. MÖLLER

Theoretical Division, Los Alamos National Laboratory, Los Alamos, New Mexico 87545, USA, E-mail: moller@lanl.gov
http://t16web.lanl.gov/Moller/abstracts.html

J. RANDRUP

Nuclear Science Division, Lawrence Berkeley National Laboratory, Berkeley, California 94720, USA, E-mail: jrandrup@lbl.gov

We study further a recent model for fission-fragment yield distributions based on Brownian shape motion on 5D potential-energy surfaces. Previously it was shown that this model describes well the transition between symmetric and asymmetric fission in the light Th region; here we study this transition near ^{258}Fm and compare to scission-type yield models. We also study the impact of the relative density of grid points in the different shape coordinates. Although extreme changes in grid spacing affect the calculated yields to some degree, we find that the full 5D model with our origina grid choice is remarkably robust and that it is therefore suitable for applications, for example fission yields relevant for fission recycling in the r-process and its termination by fission.

Keywords: Fission yields, Brownian motion.

1. Introduction

In a recent paper[1] Randrup and collaborator introduced a highly accurate method for calculating fission fragment yields at energies above the barrier based on Brownian shape motion (BSM) on 5D potential-energy surfaces. This method allows a fission yield distribution for a specific energy to be calculated on a single CPU in minutes to hours. For full specification of the model we refer to Refs.[1,2] Here we benchmark the model further and implement an improved level-density model.

2. Results

We perform three investigations: (1) effect of semi-microscopic level density, (2) yield asymmetry in the ^{258}Fm region, and (3) effect of drastic changes in grid-point spacing. Additional benchmarking is in Ref.[3]

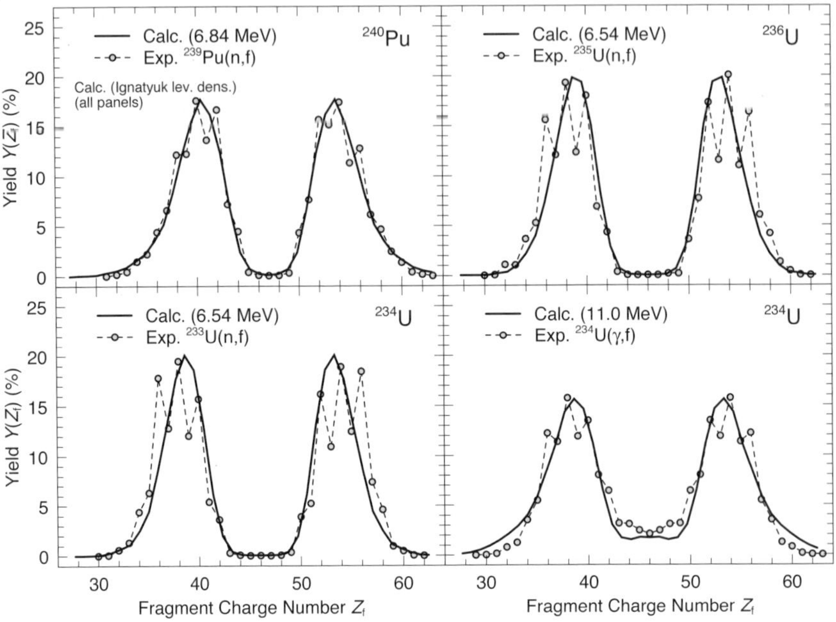

Fig. 1. Experimental yields and calculations with a semi-microscopic level density. We obtain a stronger energy dependence in the symmetric valley area than Ref.,[3] because as candidate neighbour points we use the 242 points on the 5D hypercube, Ref.[3] uses the 10 points on the coordinate axes. Our choice leads to scission in fewer steps.

2.1. *Level-density parameter*

Relative to the model implemented in[1] we have introduced a level density-parameter $a_{\rm mic}$ that takes into account the effects of variation in single-particle level structure on the level density through the level-density parameter $a_{\rm mic}$. Following Ignatyuk[4] we write

$$a_{\rm mic} = a_{\rm mac}\left\{1 + \left[1 - \exp\left(-\frac{E^*}{E_{\rm damp}}\right)\right]\frac{E_{\rm mic}(Q_2, d, \epsilon_{\rm f1}, \epsilon_{\rm f2}, \alpha)}{E^*}\right\} \qquad (1)$$

The parameters are: macroscopic level density parameter $a_{\rm mac} = A/E_{\rm l}$ with $E_{\rm l} = 8$ MeV and damping range $E_{\rm damp} = 18.5$ MeV. The shell-plus-pairing correction $E_{\rm mic}(Q_2, d, \epsilon_{\rm f1}, \epsilon_{\rm f2}, \alpha)$ is calculated for the deformed shape corresponding to each grid point from calculated folded-Yukawa single-particle levels by use of Strutinsky's method. In Fig. 1 we show our standard Pu/U test suite and find excellent agreement with experiment. In particular we find that the energy dependence is now much improved (compare the (γ,f) reaction in panel four in Fig. 1 with the corresponding result in Ref.[1]

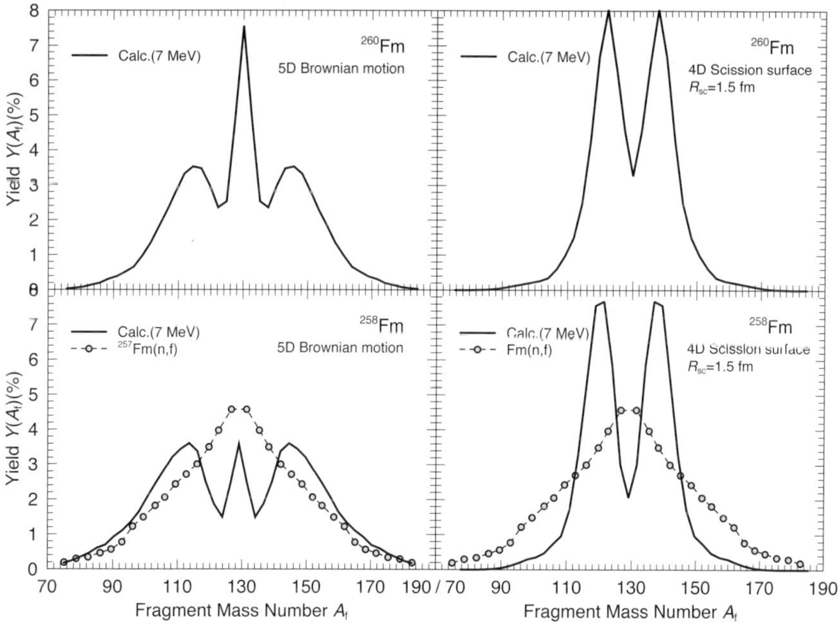

Fig. 2. Yields for ^{258}Fm and ^{260}Fm. Experimental data exist only for ^{258}Fm. The BSM model reproduces well the sharp evolution towards symmetry at and above ^{258}Fm, the scission model much less so.

2.2. *Heavy Fm region and scission models*

In many scission models[5,6] it is assumed that the yield corresponds to thermal equilibrium on a scission surface. Thus, in point i we have

$$Y_i(E^*) \approx \rho(E^*) = \exp(2\sqrt{a(E^* - E_{\text{pot}}(Q_2, d, \epsilon_{\text{f}1}, \epsilon_{\text{f}2}, \alpha))}) \tag{2}$$

Previous studies have described the scission surface in a crude, non-obvious fashion in a low-dimensional deformation space as an object consisting of relatively few points. In contrast, we define a scission surface as a 4D object, on which the 5 shape coordinates vary smoothly from grid-point to grid-point. We start by defining a neck radius r_{sci} corresponding to scission. Next, we label each point in our 5D space +1 if its neck radius is larger than r_{sci} and -1 if it is smaller. Any point labeled +1 which is next to a point labeled -1 is defined as a scission-surface point. Consequently our scission surfaces consist of more than 100 000 points. In each grid point the unnormalized yield is calculated by use of Eq. 2. We obtain the normalized charge-yield curve by carrying out the proper summations and normaliza-

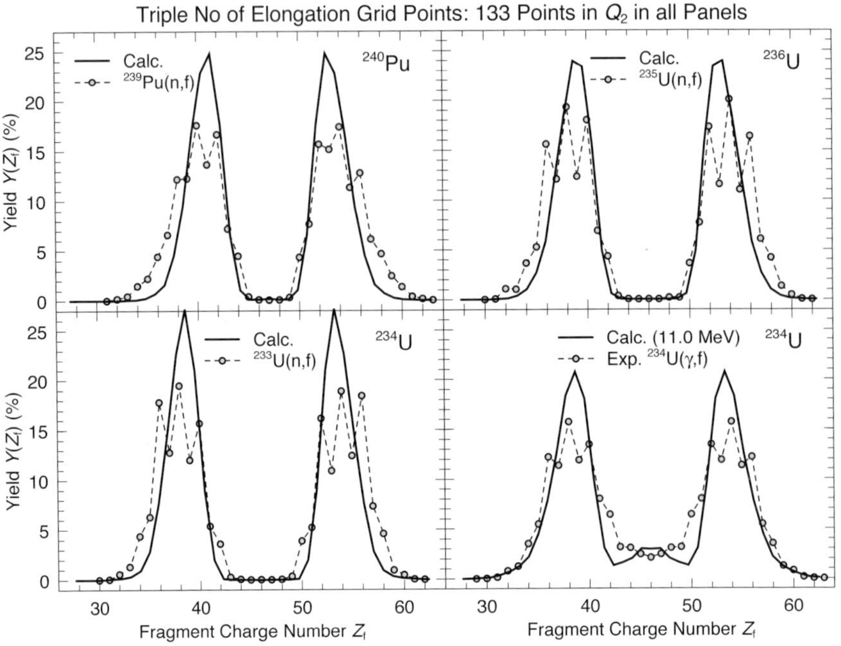

Fig. 3. Yields based on a potential energy with 3 times denser spacing of grid points in the Q_2 variable than our normal, preferred surface. Some narrowing of the distributions is observed. However, the exellent descriptions of (1) the varying width of the symmetric valley and (2) the energy dependence of the symmetric yield are remarkably robust.

tions. In Fig. 2 we show calculated yields in a BSM 5D model (our model of choice) and the above scission model for ^{258}Fm and ^{260}Fm.

2.3. *Effect of grid-point spacing*

By interpolation we have generated potential-energy surfaces that between the original Q_2 grid points have 2 additional points, effectively spacing the grid points three times denser in the fission (Q_2) direction. The calculated yields are displayed in Fig. 3. Some effect of this extreme change in grid-point spacing is seen. To make the results invariant to grid-point spacing we have calculated the overlap between a current point and its neighbor points throughout the grid. The next step (in progress) is to introduce a sampling probability that depends appropriately on the degree of overlap.

3. Conclusions

We have here and elsewhere[1,3] shown the high accuracy with which the BSM model describes fission yields, including the transitions between symmetric and asymmetric yield in the $^{220-230}$Th and ^{258}Fm regions. Although there is some dependence on the relative number of grid points in the five shape coordinates, see Fig. 3, we find that even under rather extreme changes the results are fairly robust. Clearly the number of grid points in the elongation direction Q_2 relative to the other shape variables introduces implicitly a time-scale of fission. However, it is a remarkable discovery that we obtain such robust results in a model whose specification took no account of fission yields. The calculated potential energy must therefore be highly realistic and further studies hold promise of new insights into the nature of the fission process.

An advantage of the method that cannot be overstated, is that a yield can be obtained in minutes to hours on a single CPU once the 5D potential energy is calculated. We have such surfaces calculated between the proton and neutron drip lines for all 5254 nuclides in the range $170 < A \leq 330$. Therefore the approach can immediately be used to calculate yield data bases for fission relevant to r-process modeling.

This work was supported by travel grants for P. M. to JUSTIPEN (Japan-U. S. Theory Institute for Physics with Exotic Nuclei) under grant number DE-FG02-06ER41407 (U. Tennessee) and carried out under the auspices of the National Nuclear Security Administration of the U. S. Department of Energy at Los Alamos National Laboratory under Contract No. DE-AC52-06NA25396 (PM) and at the Lawrence Berkeley Laboratory under Contract No. DE-AC02-05CH11231 (JR). This work also benefitted from discussions with T. Ichikawa, A. Iwamoto, A. J. Sierk, and J. Lestone.

References

1. J. Randrup and P. Möller, Phys. Rev. Lett. **106** (2011) 132503.
2. P. Möller, D. G. Madland, A. J. Sierk, and A. Iwamoto, Nature **409** (2001) 785.
3. J. Randrup, P. Möller, and A. J. Sierk, submitted to Phys. Rev. C, arXiv:1107:2624 [nucl-th].
4. A. V. Ignatyuk, G. V. Smirenkin, and A. N. Tishin, Yad. Fiz. **21** (1975) 485.
5. P. Fong, Phys. Rev. **102** (1956) 434.
6. B. D. Wilkins, E. P. Steinberg, and R. R. Chasman, Phys. Rev. C **14** (1976) 1832.

Digital electronics for γ-ray spectroscopy - the path to low cross-sections : First evidence of a rotationnal band in ^{246}Fm

Piot J. for the JR83 experiment collaborators
Institut Pluridisciplinaire Hubert Curien, 23 rue du Loess, 67037 Strasbourg, France

We present the installation and validation of the TNT2D digital cards on the JUROGAM and JUROGAM2 germanium arrays and their use for the low cross-section experiment on ^{246}Fm.

Keywords: Gamma-ray, spectroscopy, Digital signal processing, TNT2D, JUROGAM2, Fermium, Fusion-evaporation

1. Prompt gamma-ray spectroscopy for low cross-sections

The region of transfermium nuclei is characterized by a sharp decrease of the fusion-evaporation cross-section. Higher beam intensities are therefore required to keep investigating this region with statistics compatible with spectroscopy. This means an increase of the counting rate in germanium detectors placed around the target. In order to cope with these counting rates, digital acquisition cards are needed to replace the standard analogue ADC used for γ-ray spectrosocpy. The TNT2D cards (Tracking Numerical Treatement 2 Dubna)[1] have been developpped by the team of P. Medina at IPHC to fit the needs of γ-ray spectroscopy with high counting rates and Compton-scattering discrimination. These acquisition cards have been installed and validated on the JUROGAM Germanium array at the University of Jyvskyl. Their use for the prompt γ-ray spectroscopy of low cross-section nuclei shows the improvements brought by the use of digital electronics for these experiments.

2. The TNT2D digital cards

The TNT2D are four channel digital acquisition cards sampling the signal at 100 MHz with a 14 bits flash ADC. The cards are designed for gamma-ray spectroscopy with the use of anti-Compton BGO shields. Signals coming from the preamplifiers are shaped online into a trapezoid[2] from which is deduced the energy measurement. This processing considerably reduces the deadtime and improves the energy resolution. For further details regarding the TNT2D cards, please refer to.[1]

Before using the TNT2D cards as the acquisition system for JUROGAM, they have been compared to the existing analogue acquisition during an experiment. The detectors were connected to both the analogue ADCs and the TNT2D cards whose data were recorded in parallel.[3]

Study of the energy linearity

The linearity of the acquisition system is critical for γ-ray spectroscopy. In order to estimate the linearity of both the analogue and digital acquisition systems, their records of a calibration run using a ^{152}Eu - ^{133}Ba sources were compared to the reference values of the transitions[4] (see Fig. 1). The deviation for the TNT2D cards stays below 0.5 keV, whereas it reaches 3 keV for analogue ADCs. This implies an integral non-linearity of 0.0065% for the TNT2D cards compared to 0.022% for the analogue ADCs.

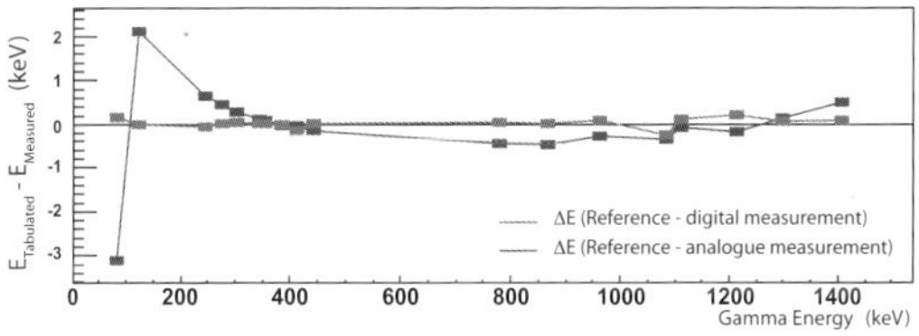

Fig. 1. Energy linearity for both analogue and digital acquisition

Correlation between the digital and analogue data

The γ-ray events from both data sets were compared using the synchronisation signals from the absolute clock of the Total Data Readout system to synchronise the data files. The γ-ray events were then compared based on the timestamps, the channel numbers, and the energies. The time window for the identification was 500 ns wide. The results show 88% correlated events between both data sets. The digital cards bring a 36% increase in the statistics compared to the analogue recorded data for an average counting rate of 10 kHz.

Automated offset for clover detectors

The crystals of the clover detectors are coupled to their preamplifier by a capacitive circuit. This coupling filters out the slow component of the signal coming from the detector. The closer and higher the signals, the more the baseline of the ADC will diverge from the original value. An offset is defined in the TNT2D settings to fit the baseline and the signal amplitude within the dynamic of the ADC. However, if the baseline shifts with respect to the counting rate, part of the signal is lost due to saturation. This has been solved by adding an automatic offset adaptation. Threshold values of the counting rate are defined and associated with a fitting offset value. The card then samples the counting rate and adapts the offset value accordingly. An hysteresis cycle is used to avoid oscillations around the threshold values.

3. ^{246}Fm : Prompt gamma-ray spectroscopy at the limits

The use of digital electronics allowed to measure the prompt deexcitation of nuclei ^{246}Fm with a record fusion-evaporation cross-section of 11 nb. The experiment was performed at the University of Jyvskyl using the JU-ROGAM2 array associated with the magnetic separator RITU[5] and the focal plane detection system GREAT.[6] The ^{246}Fm nuclei were produced through the reaction of fusion evaporation ^{208}Pb(^{40}Ar,2n)^{246}Fm.

The experiment yielded 276 alpha particles identified as the decay of ^{246}Fm through their energy and decay time.[7] The correlation of implanted nuclei with their alpha decay provided a first unambiguous prompt gamma-ray spectrum showing the presence of a rotational structure. This correlation also provided the position of ^{246}Fm nuclei in the implantation energy *vs* time of flight (E-Tof) matrix. Using a selection based on this position, a wider selection was obtained for the nuclei of interest. This method yielded the prompt gamma-ray spectrum shown in figure 2. The labeled transitions belong to the rotational structure previously observed.

The dynamical moment of inertia shows a behaviour similar to the other known even-even isotopes of Fermium (^{248}Fm[9] and ^{250}Fm[10]). The fit of the dynamical moment of inertia using the Harris formula[11] provided the spin values corresponding to the parent states of the observed transitions. The extrapolation of this fit to low frequencies gave the energies of the two first transitions of the band, unseen during the experiment because of their high electron conversion probability. The level scheme is shown on figure 2.

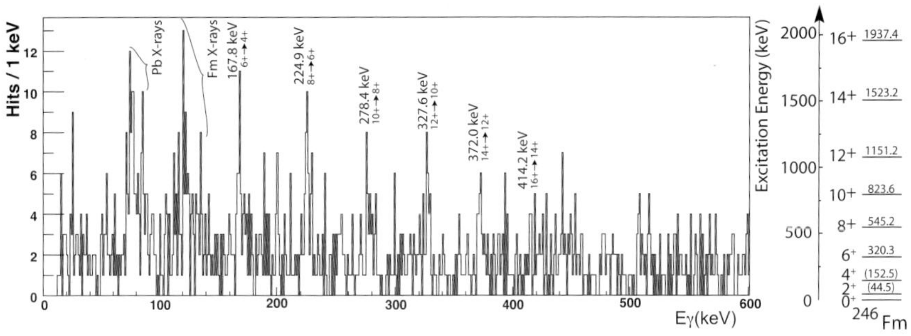

Fig. 2. Left: Prompt gamma-ray spectrum of ^{246}Fm. Right: Level scheme of the ground-state based rotationnal band of ^{246}Fm.

Summary and conclusion

The installation of the TNT2D digital cards on the JUROGAM2 array allowed to record the prompt gamma-ray de-excitation of ^{246}Fm with a production cross-section of 11 nb. This experiment marks a milestone for gamma-ray spectroscopy. The use of digital electronics pushes the limits of gamma-ray spectroscopy, allowing to cope with much higher counting rates than previously. The spectroscopy of nuclei in the 10 nb region and lower is now accessible.

Acknowledgments

The author thanks P. Medina, M. Richer, L. Charles and C. Santos for their hard work on the TNT2D cards. The author thanks the member of the ^{246}Fm experiment at JYFL, especially B. Gall, O. Dorvaux, P.T. Greenlees, P. Jones and T.-L. Khoo, for their precious help.

References

1. Arnold L. *et al.*, IEEE Trans. Nucl. Sci. 53 (2006), 723.
2. Jordanov V.T. and Knoll G.F., Nucl. Inst. Meth. A 345 (1994), 337.
3. Piot .J *et al.*, to be published.
4. NNDC, http://www.nndc.bnl.gov/nudat2/chartNuc.jsp
5. Leino M. *et al.*, Nucl. Inst. Meth. B 99 (1995), 653.
6. R.D. Page *et al.*, Nucl. Inst. Meth. B 204 (2003) 634.
7. M. Venhart *et al.*, Eur. Phys. Jour. A 47 (2009) 20.
8. Piot J. *et al.*, to be published.
9. Herzberg R.-D. and Greenlees P.T., Prog. Part. Nucl. Phys., 61 (2007) 674.
10. Greenlees P.T. *et al.*, Phys. Rev. C 78 (2008) 021303.
11. Siem, S. *et al.*, Phys. Rev. C 70 (2004), 014303.

MINOS: a vertex tracker for in-beam γ spectroscopy at relativistic energies.

A. Obertelli[†] on behalf of the MINOS collaboration

CEA, Centre de Saclay, IRFU/Service de Physique Nucléaire, F-91191 Gif-sur-Yvette, France

[†] *E-mail:alexandre.obertelli@cea.fr*

In-flight gamma spectroscopy of rare isotopes at relativistic energies, from fifty to several hundreds of MeV/nucleon, is one of the most efficient tools to investigate shell effects in exotic nuclei. We propose a new method to increase the sensitivity of prompt-gamma spectroscopy by more than one order of magnitude compared to existing setups, leading to a significant step forward. The intended program is based on proton-induced knockout reactions such as $(p, 2p)$. Experiments should take advantage of the most exotic neutron-rich beams produced at the RIBF[1] in RIKEN and the upcoming european FAIR facility. MINOS has been funded by the European Research Council for the 2010-2015 period.

Keywords: Detector development; Hydrogen target ; Micromegas; Gamma spectroscopy

1. General description

MINOS, quasi-acronyme for "MagIc Numbers Off Stability", targets the spectroscopy of the most neutron-rich nuclei produced at fragmentation facilities. It consists in a new technique based on nucleon-removal from very exotic nuclei on a very thick liquid hydrogen target coupled to a proton tracker that aims at measuring, on an event-by-event basis, the reaction-vertex position inside the target. By measuring the reaction vertex, one can use targets of hundreds of millimeters while improving the gamma-detection sensitivity, *i.e.* the Doppler correction is better than with a standard heavy-ion target. The only remaining limiting factor is to ensure that the second-interaction probability in the target is low in order to be able to identify the reaction residue after the target. For incident energies of E=200-500 MeV/nucleon, a typical length of 150 mm allows to respect this condition. This development will induce a unique gain in detection sensitivity of more than an order of magnitude compared to experiments with solid heavy-ion

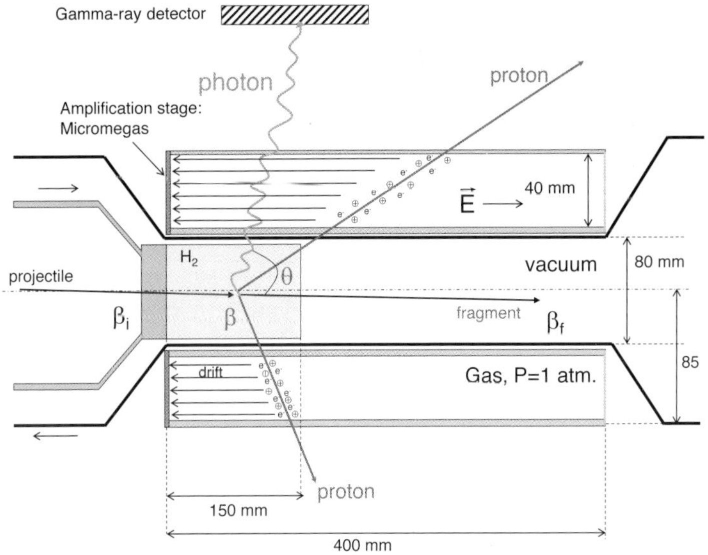

Fig. 1. Conceptual scheme of a target-TPC device for the MINOS project. The extracted vertex position is projected to be inferior to 3 mm FWHM. The tracker measures with precision the emission angle θ and the velocity β of the fragment at the vertex position, essential for a good Doppler correction.

targets, allowing the detailed spectroscopy of nuclei produced at less than 1 particle per second. The technique will permit exclusive measurements and the selection of kinematics conditions for a direct reaction mechanism and a proper spectroscopic factor extraction.

A pure liquid H_2 target, when used alone, still keeps some limitation due to energy loss in the target and its spatial extension which, if larger than a centimeter, limits the angular resolution for the Doppler correction. In case of a thick target of 150 mm, the velocity spread is about 8-20% for an initial velocity corresponding to 200-500 MeV/nucleon, depending on the considered incident ion. This example shows that such a thick target cannot be used for high-resolution gamma spectroscopy if the vertex position is not known. This is one of the motivations for the MINOS project that will couple a thick cryogenic LH_2 target to a charged-particle tracker in order to reconstruct the vertex position. The development of such a device for intermediate-energy (p,Xp)-like reactions coupled to an hydrogen target is possible since (i) at least one light-charged particle (proton) is emitted during the reaction, (ii) the energy loss of these protons in H_2 is most of the time small compared to their kinetic energy. The objective is to reach a 3-mm vertex-position resolution FWHM. In the case of a $(p, 2p)$ measurement such as ^{63}V(p,2p)^{62}Ti at 250 MeV/nucleon and a target thickness of 150-mm, the energy loss in the target represents 30% of the total kinetic energy, i.e. a $\Delta\beta$=9% velocity spread from β=0.61 to β=0.53 along the target thickness. A Δx=3-mm resolution in the vertex position gives a $\Delta\beta/\beta$ =0.2% velocity resolution and does not impact the angular resolution for the scattering angle needed in the Doppler-effect correction. An equivalent energy resolution is found in a 3-mm thick ^{9}Be target, with approximately 20 times less atoms.cm^{-2}. Considering that deeply-bound proton removal reactions at intermediate energies have similar cross sections from H, ^{9}Be or ^{12}C, this leads to a net gain of a factor ~20 in statistics.

The device and its readout system will be conceived and built at CEA Saclay DSM/IRFU, France. The target will have a similar design (a finger-shaped envelope) as the recent development[2] for the PRESPEC collaboration. The geometry and material of the target will be chosen to minimize the absorption at all angles (including 90 degrees in the laboratory frame). The angular straggling in hydrogen for such a thick target remains rather low. SRIM Monte-Carlo simulations for ^{110}Zr at 350 MeV/nucleon give an angular straggling of approximately 3 mrad FWHM.
The trajectories of protons will be analyzed using a cylindrical time-projection chamber (TPC). A tracker solution instead of a TPC is also under study at the present stage. The TPC will surround the target and be 300-400 mm long to access the detection of particles emitted at small angles (see a schematic view in Fig. 1), will detect hadrons in the momentum range from 300 to 1500 MeV/c. In the case of (p, pn) reactions, it corresponds to the detection of recoil protons whereas in the case of knockout $(p, 2p)$ reactions both the recoil and the projectile-like protons will be detected in coincidence. Angular resolution should be of the order of 20 mrad or less. Drift electrons will be reaching the backplane and amplified via a Micromegas stage, as the one developped by CEA Saclay DSM/IRFU for the readout of the T2K/TPCs.[3,4] Charges will be induced on a pixelized plane with pads of the order of 2 to 4 mm^2 for a total amount of the order of 5000 channels. Signals will be digitized via a specific electronics and readout possibly based on the GET (General Electronics for TPCs) developments,[5] a collaboration between CEA-IRFU, IN2P3 and MSU/NSCL.

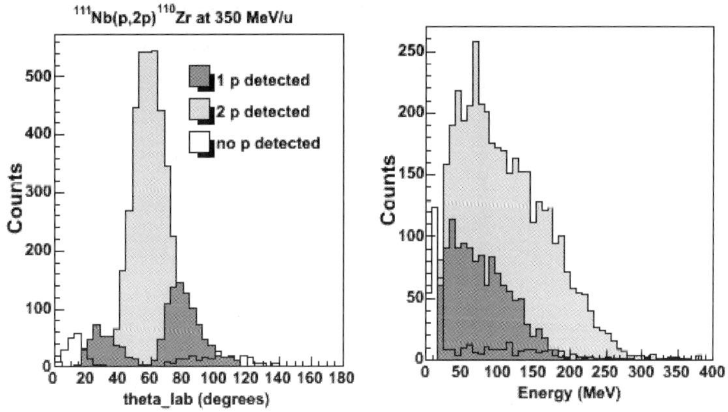

Fig. 2. Angular (left panel) and energy (right panel) distributions for protons emitted from the ^{111}Nb(p,2p)^{110}Zr reaction at 350 MeV/nucleon. Events where no proton is detected (white), one proton (blue) or two protons (pink) are shown.

2. Expected performance

We performed preliminary simulations of the tracker with realistic events from Monte-Carlo simulations (see Fig. 2). Events have been generated with an intra-nuclear cascade code. The target radius is assumed to be 35 mm and the internal and external radius of the TPC were 45 and 85 millimeters, respectively. The following conclusions have been obtained from the two systems ^{110}Zr+p and ^{54}Ca+p at 350 MeV/nucleon. Energy losses and detection thresholds have been considered. The detection efficiency for $(p, 2p)$ reactions is larger than 85% with approximately 53% of events with both protons detected and 32% where only one is detected. In case of $(p, 3p)$ reactions, the detection efficiency is even higher since the exit channel contains more protons. Interestingly, (p, pn) reactions are also detected with a significant efficiency of more than 70%.

Simulations of the gamma detection system with the GEANT4 software have been performed for the same reactions to characterize the gain in sensitivity expected from this development (see Fig. 3). We considered the population of ^{54}Ca(2^+) and ^{110}Zr(2^+) via one-nucleon knockout. Both transitions has been considered as very short lived. We considered in each case (i) the foreseen development with a vertex position resolution of 3-mm FWHM and (ii) a "standard" ^{9}Be target with a thickness limiting the velocity spread in the target to 5% FWHM. The statistics corresponds to one

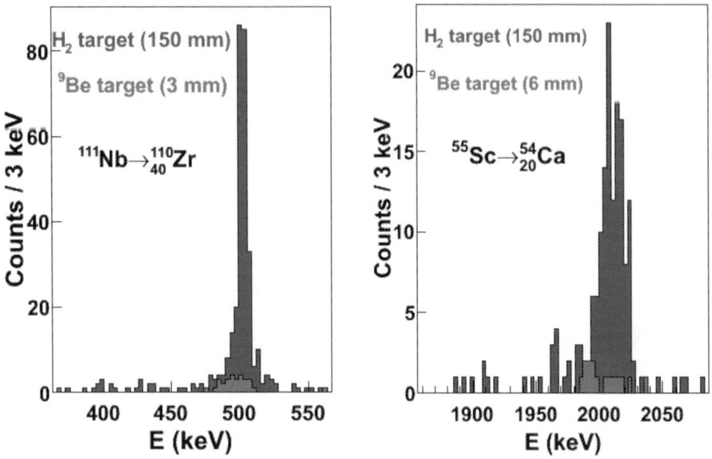

Fig. 3. Simulated gamma spectra obtained for ^{111}Nb(p,2p)^{110}Zr and ^{55}Sc(p,2p)^{54}Ca at 350 MeV/nucleon.

week of beam time for a one-nucleon knockout cross section of 10 mb and a beam intensity of 1 particle per second. Simulations have been performed with the AGATA array with a one-π configuration at forward angles located at 20 centimeters from the center of the target in each simulation. Simulations show that the figure of merit, taken as proportional to the ratio of the statistics to the energy resolution, reaches 15 in the case of ^{54}Ca and 30 in the case of ^{110}Zr. Similar conclusions can be raised for incident energies down to $\sim$ 200 MeV/nucleon and a 100-mm thick target.

The effect of the lifetime of the populated states on the line shape of the measured photopeak has been estimated through simulations. A distorsion of the lineshape is visible for lifetimes larger than 20 picoseconds (ps). For lifetimes from 20 ps to 100 ps, the shape of the measured photopeak can be used as a measurement of the lifetime with still a significant gain compared to the use of a thin target. In the case of a longer lifetime, the distorsion of the photopeak starts to be too large to maintain a good-enough sensitivity.

3. Foreseen agenda

The cryogenic target and Time-Projection Chamber developments will be undertaking through to 2012. The full device is expected to be complete in 2013. First radioactive-beam experiments could take place at the

RIBF/RIKEN as soon as 2013.The architecture of the system will be defined to fit inside the DALI2 γ spectrometer and to be used either with the zero-degree spectrometer[6] or with the SAMURAI setup.[7]

This work is supported by the European Research Council (ERC) through the Starting Grant 258567-MINOS.

References

1. Y. Yano, Nucl. Instr. Res. Meth. B **261**, 1009 (2007).
2. A. Obertelli and T. Uesaka, to be published in Eur. Phys. J. A (2011).
3. Y. Giomataris *et al.*, Nucl. Instr. Res. Meth. A **376**, 29 (1996).
4. N. Abgrall *et al.*, Nucl. Instr. Res. Meth. A **637**, 25 (2011).
5. E. Pollacco, French national Grant ANR-09-BLAN-0203 (2009-2013).
6. T. Kubo, Nucl. Instr. Res. Meth. B **204**, 97 (2003).
7. K. Yoneda *et al.*, RIKEN Accel. Prog. Rep. **43**, 178 (2010).

Elastic scattering of protons with radioactive ion beams –Overview of ESPRI project–

J. Zenihiro[a], Y. Matsuda[b], H. Sakaguchi[c], S. Terashima[d], H. Otsu[a], H. Takeda[a], K. Ozeki[a], K. Yoneda[a], K. Tanaka[a], T. Ohnishi[a], T. Kobayashi[b], T. Murakami[e], Y. Maeda[f], Y. Sato[g], I. Tanihata[c], O. H. Jing[c], M. Takechi[d], and M. Kanazawa[i]

[a]*RIKEN Nishina Center, Saitama 351-0198, Japan*
[b]*Department of Physics, Tohoku University, Miyagi 980-8578, Japan*
[c]*Research Center for Nuclear Physics, Osaka University, Osaka 567-0047, Japan*
[d]*Gesellshaft fur Schwerionenforschung, D-64291 Darmstadt, Germany*
[e]*Department of Physics, Kyoto University, Kyoto 606-8502, Japan*
[f]*Department of Applied Physics, University of Miyazaki, Miyazaki 889-2192, Japan*
[g]*Department of Physics and Astronomy, Seoul University, Seoul 151-747, Korea*
[h]*National Institute of Radiological Sciences, Chiba 263-8555, Japan*

A new experimental apparatus has been developed to measure proton elastic scattering from radioactive ion beams using inverse kinematics. We have performed experiments for several unstable nuclei such as ^{9,10,11}C, ^{20}O, and ^{66,70}Ni. While the experiments, the new apparatus has provided a high performance of a long-term use of 1-mm-thick and 30-mm-ϕ solid hydrogen target, and a excitation energy resolution of $\Delta E_x \sim 500$ keV.

Keywords: nucleon density distribution, equation of state, unstable nuclei, proton elastic scattering, solid hydrogen target

1. Introduction

Nucleon density distribution is one of the most fundamental properties of nuclei. In the case of stable nuclei, various nuclear charge distributions are now known accurately from a large number of experiments such as electron scattering, muonic X-ray, and isotope shift. Proton density distributions can be derived from the nuclear charge distributions.[1] Recently we have succeeded in extracting neutron density distributions of stable nuclei from proton elastic scattering at intermediate energy $E_p = 300$ MeV.[2–4]

On the other hand, the information about unstable nuclei is very few because of the experimental difficulties. In recent years, the nuclear many-body system with large isospin asymmetry has been studied in both the

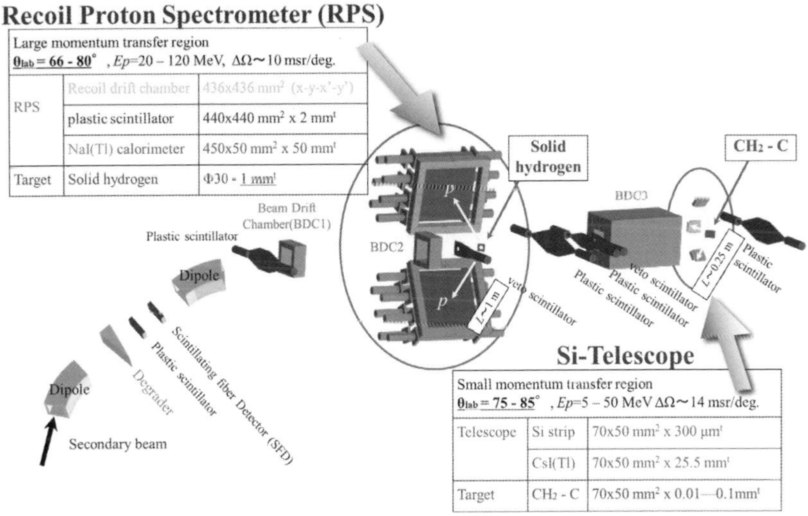

Fig. 1. Schematic 3D view of typical setup for ESPRI measurements and specifications of RPS and Si-telescope.

nuclear physics and astrophysics.[5] In particular, the isospin dependent term of the equation of state (EOS) is very important to understand the astrophysical phenomena such as a supernova explosion, a neutron star structure or cooling system. Therefore, the neutron and proton density distributions of unstable nuclei with large isospin asymmetry are very powerful tools to determine the EOS of asymmetric nuclear matter.

If we could measure proton elastic scattering of unstable nuclei, we would apply our analysis method to unstable nuclei as well as stable nuclei. RI beam factory (RIBF) in RIKEN is the one of the next generation facilities in the world, and has already started providing heavy-ion beams such as ^{238}U at 345 MeV/A.

2. Setup for ESPRI measurements

We have proposed a new experimental project to extract the nucleon density distributions of proton- or neutron-rich exotic nuclei by means of Elastic Scattering of Protons with Radioactive Ion beams (ESPRI) at 300 MeV/A. In ESPRI experiments, recoil protons are detected in inverse kinematics and angular distributions of cross sections are measured. In the case of unstable nuclei, we need to use inverse kinematics because we cannot use unstable nuclei as a fixed target. Protons, which are fixed in the target position, are inverse-kinematically scattered by heavy ion beams of unstable nuclei. We have planned to use missing mass spectroscopy technique to obtain the elastic scattering data.

First, we have developed a new experimental apparatus to detect four momenta of recoil protons over a wide range of momentum transfer from 1.0 fm^{-1} to 2.5 fm^{-1}. Since recoil proton energy and cross section change drastically with an increase of the momentum transfer, it has been difficult to measure the scattering in such an extensive region of momentum transfer. The apparatus for ESPRI experiments is divided into two parts. One is called recoil proton spectrometer (RPS) which covers a wide range of recoil angle ($66^\circ \leq \theta_{lab} \leq 80^\circ$). RPS consists of a solid hydrogen target (SHT), two plastic scintillators, two multi-wire-drift chambers (MWDCs), and 14 NaI(Tl) calorimeter arrays. The flight length from SHT to MWDCs is about 1 m. The other is called Si-telescope which consists of three Si-strip + CsI(Tl) sets to measure recoil protons in the small momentum transfer region around 1 fm^{-1}. Figure 1 shows a typical setup for ESPRI measurements, which consists of RPS, Si-telescope, and beam line detectors. Specifications of RPS and Si-telescope are also listed in Fig. 1.

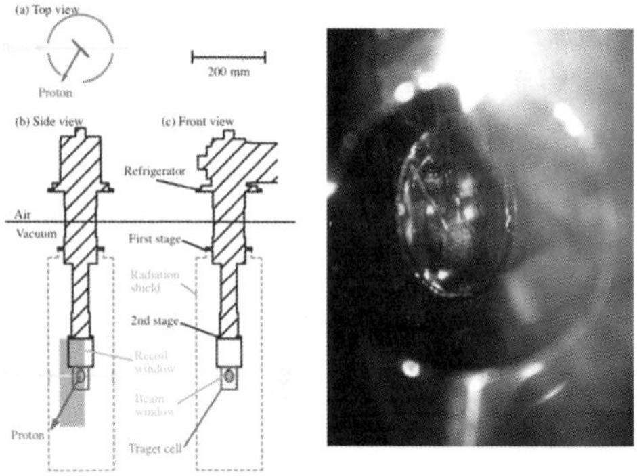

Fig. 2. Schematic view of the cryogenic system of SHT (left side); (a) top, (b) side, and (c) front views. Picture of SHT (30-mm-ϕ and 1-mm-thick) which was used during a measurement (right side).

A specific equipment of this system is the solid hydrogen target of RPS. The excitation energy resolution is mainly governed by the angular resolution. In order to get a good angular resolution, the target thickness must be thin enough to reduce the multiple scattering in the target material.

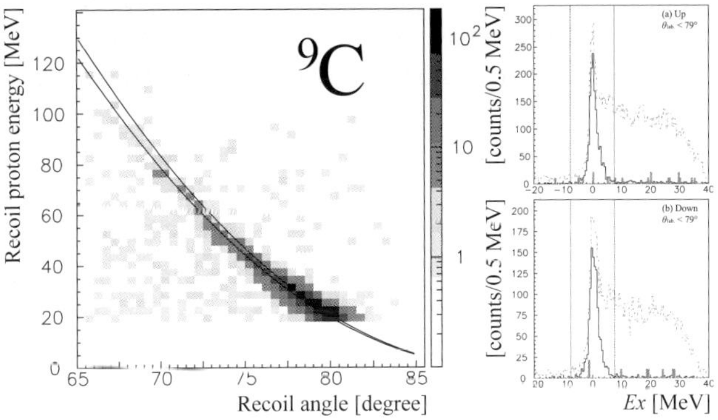

Fig. 3. Kinematics correlation between recoil proton energy and angle (left side). Excitation energy spectra of H(^{9}C, p)^{9}C reaction measured by upper and lower part of RPS, respectively (right side).

However, meaningful statistics are also required for this measurement. The number of protons in SHT is ten times larger than that of a polyethylene target which gives the same multiple scattering spread. In addition, carbon in polyethylene causes many background events. Recently we have succeeded in making 1-mm-thick and 30-mm-diameter SHT.[6] Due to the thermal radiation problem it has been difficult to make such a thin and large SHT, so far. Thermal radiation is mainly caused by heat flow from surroundings and small thermal conductivity of normal hydrogen. To avoid this thermal problem, we have used almost 100% *para*-hydrogen whose thermal conductivity at low temperature is 100 times larger than that of normal hydrogen, which contains only about 25% *para*-hydrogen. At low temperature, the conversion from *ortho*-state to *para*-state gradually occurs, but it takes very long time without a catalyst. Thus, we have used an *ortho-para* converter with Iron(III) oxide-hydroxide catalysis to enhance the conversion rate. Schematic view of the cryogenic system of SHT is shown in the left side of Fig. 2. We have finally obtained the stable SHT as shown in the right side of Fig. 2.

3. Experiment

After developing ESPRI detectors as mentioned above, we have performed experiments for several unstable nuclei. ESPRI experiments for light unstable nuclei of ^{9,10,11}C and ^{20}O were performed at heavy ion medical accelerator in Chiba (HIMAC), and for medium-heavy nuclei of ^{66,70}Ni at GSI, Germany. Analyses are now on progress, thus in this report we show some preliminary results of excitation energy spectra. The systematic changes of the nucleon densities of light and medium-heavy isotopes will help us to understand the changes of single particle states depending on the isospin asymmetry, and to determine the nuclear matter EOS, respectively.

In the case of ^{9}C, for example, ^{12}C primary beam, which was accelerated to 430 MeV/A by the synchrotron-type accelerator, bombarded ^{9}Be production target. After the production target, ^{9}C beam with an energy of 290 MeV/A was selected with a fragment separator and the beam line detectors. Finally, ^{9}C secondary beam reached the SHT or polyethylene target position. The H(^{9}C, p)^{9}C reaction was reconstructed event by event with beam line and RPS and Si-telescope detectors. Since ^{9}C has no bound exited states, we used 5-mm-thick SHT in this measurement in order to increase the yield. The left side of Fig. 3 shows the correlation plot between energy and angle of recoil protons via the H(^{9}C, p)^{9}C reaction. Excitation spectra by upper and lower part of RPS are obtained and the ground state peaks ($\Delta E_x \sim 1$ MeV) are clearly seen in the right side of Fig. 3.

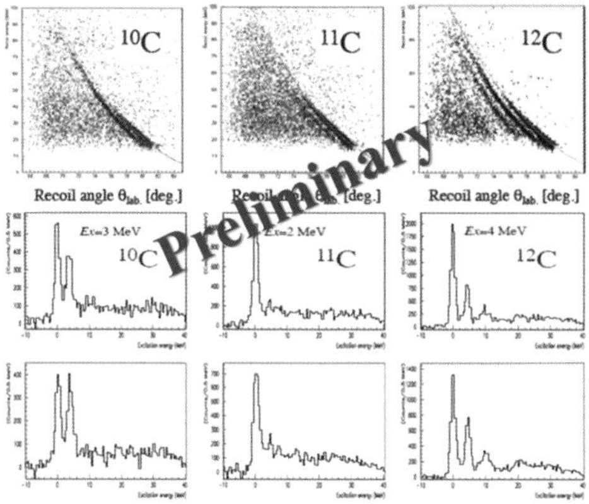

Fig. 4. Similar to Fig. 3, but the results of ^{10,11,12}C in the left, middle, and right, respectively.

Figure 4 shows the preliminary results of H(^{10,11,12}C, p) measurements. Different from ^{9}C case, we have used 1 mm thick SHT because ^{10,11,12}C have bound excited states. The obtained excitation spectra show much better energy resolutions of $\Delta E_x \sim 500$ keV than that of ^{9}C case. In this report, very preliminary results of ^{20}O and ^{58,66,70}Ni data are not shown, but they indicate the similar level of the excitation energy resolution to

^{10,11,12}C case. Throughout the measurements the yield and appearance of 1-mm-thick SHT has not changed at all. Thus, we have confirmed that the 1-mm-thick SHT can maintain its thickness and solid state for long-term irradiation of radioactive beam. Now we have achieved the good enough performance of the ESPRI apparatus to measure the proton elastic scattering from radioactive beams.

4. Summary

We have proposed a so-called ESPRI project to deduce the nucleon density distributions of exotic nuclei with large isospin asymmetry from the measurement of proton elastic scattering at $E_p = 300$ MeV/A. We have developed a new apparatus for the ESPRI experiments, which covers extensive momentum transfer region from 1 fm^{-1} to 2.5 fm^{-1}, and succeeded in making and maintaining the large and thin SHT target (30 mm in diameter and 1 mm thick) using *para*-hydrogen gas. Experiments for several unstable nuclei (^{9,10,11,12}C, ^{20}O, ^{66,70}Ni) with the ESPRI apparatus have been also performed in HIMAC, Japan and GSI, Germany. Some preliminary results of cross section data have been obtained and will be shown in near future. We now prepare for an approved ESPRI experiment to measure the proton elastic scattering from neutron-rich carbon isotopes ^{16,18}C at RIKEN-RIBF.

References

1. H. D. Vries, C. W. D. Jager and C. D. Vries, *Atomic Data and Nuclear Data Tables* **36**, 495 (1987).
2. H. Sakaguchi, H. Takeda, S. Toyama, M. Itoh, A. Yamagoshi, A. Tamii, M. Yosoi, H. Akimune, I. Daito, T. Inomata, T. Noro and Y. Hosono, *Phys. Rev. C* **57**, 1749(Apr 1998).
3. S. Terashima, H. Sakaguchi, H. Takeda, T. Ishikawa, M. Itoh, T. Kawabata, T. Murakami, M. Uchida, Y. Yasuda, M. Yosoi, J. Zenihiro, H. P. Yoshida, T. Noro, T. Ishida, S. Asaji and T. Yonemura, *Phys. Rev. C* **77**, p. 024317(Feb 2008).
4. J. Zenihiro, H. Sakaguchi, T. Murakami, M. Yosoi, Y. Yasuda, S. Terashima, Y. Iwao, H. Takeda, M. Itoh, H. P. Yoshida and M. Uchida, **82**, p. 044611(Oct 2010).
5. B.-A. Li, L.-W. Chen and C. M. Ko, *Physics Reports* **464**, 113 (2008).
6. Y. Matsuda, H. Sakaguchi, J. Zenihiro, S. Ishimoto, S. Suzuki, H. Otsu, T. Ohnishi, H. Takeda, K. Ozeki, K. Tanaka, S. Terashima, Y. Maeda, T. Kobayashi, A. Koreeda and K. Kamei, *Nuclear Instruments and Methods in Physics Research Section A: Accelerators, Spectrometers, Detectors and Associated Equipment* **643**, 6 (2011).

Studies of spin-dependent interactions in unstable nuclei with solid polarized proton target

S. Sakaguchi

Department of Physics, Kyushu University,
Hakozaki, Fukuoka, 812-8581, Japan
E-mail: sakaguchi@phys.kyushu-u.ac.jp

T. Wakui

Cyclotron and Radioisotope Center, Tohoku University,
Sendai, Miyagi, 980-8578, Japan

T. Uesaka

Nishina Center, RIKEN,
Wako, Saitama, 351-0198, Japan

The vector analyzing power has been measured for the elastic scattering of neutron-rich ^{6}He from polarized protons at 71 MeV/A making use of a newly constructed solid polarized proton target operated in a low magnetic eld of 0.1 T and at a relatively high temperature of 100 K. An optical model analysis revealed that the spin-orbit potential for ^{6}He is characterized by shallow and long-ranged shape compared with the global systematics of stable nuclei. Such a characteristics re ect a di used density distribution of the neutron-rich isotopes.

Keywords: Polarized proton; Elastic scattering; Helium isotope; Analyzing power.

1. Solid polarized proton target for RI-beam experiment

Recently, renewed interest has been focused on the manifestation of the spin-dependent interactions in unstable nuclei. One of the most powerful probes to extract the information on the spin-dependent interaction is the direct reaction induced by spin-polarized light ions. In order to investigate the unstable nuclei with spin polarization, a polarized proton solid target, which was specially designed for radioactive-ion beam experiments, has been developed at CNS, University of Tokyo.[1] The most prominent advantage of this target is in its modest operation conditions, i.e. a low magnetic eld of 0.1 T and a high temperature of 100 K. These conditions allow us to detect low energy recoiled protons, which are essential for event selection.

In order to achieve these "relaxed" conditions, a new polarizing method is employed. Firstly, electron population di erence is produced among a triplet state of photo-excited aromatic molecules.[2] Then, this electron population di erence is transferred to protons by so-called cross-polarization technique.[3] It should be noted that the magnitude of the electron population di erence is independent of the temperature and magnetic eld strength. This allows us to obtain large (20%) proton polarization in low magnetic eld and at high temperature.

2. p-^{6}He analyzing power measurement

In applying the polarized proton target to RI-beam experiments, we have started by measuring the vector analyzing power for the proton elastic scattering from neutron-rich helium isotope ^{6}He at 71 MeV/A. The aim of the measurements is to investigate the characteristics of spin-orbit potential between protons and ^{6}He, and discuss the e ect of valence neutrons on the spin-orbit potential. Since the spin-orbit coupling is essentially a surface e ect, the form factor of the spin-orbit potential is usually modeled by the radial derivative of density distribution. Thus, in neutron-rich nuclei with signi cantly di used density distribution, it is naturally expected that the shape of the spin-orbit potential can be di erent from those in stable nuclei.

Measurement of the analyzing powers for p-^{6}He elastic scattering was carried out at the RIKEN Accelerator Research Facility (RARF) in 2005 and 2007 using the RIKEN Projectile-fragment Separator (RIPS).[4] The polarized proton solid target was used as a secondary target. The target polarization during the beam-time was found to be 13.8 2.7% on average. Recoiled protons were detected by a pair of detector arrays consisting of a drift chamber (SWDCs) and a CsI (Tl) scintillator placed on the left and right sides of the beam line as shown in Fig. 1. Scattered particles were detected using a multiwire drift chamber (MWDC) and E E plastic scintillator hodoscopes.

The measured di erential cross sections and analyzing powers are shown in Fig. 2 by closed circles. The present data are consistent with those reported in Ref.[7] in the overlapping angular region. For comparison, data for the p+^{4}He[5] and p+^{6}Li[6] elastic scatterings at 72 MeV/A are also shown by open squares and triangles, respectively.

A phenomenological optical model analysis is carried out to investigate

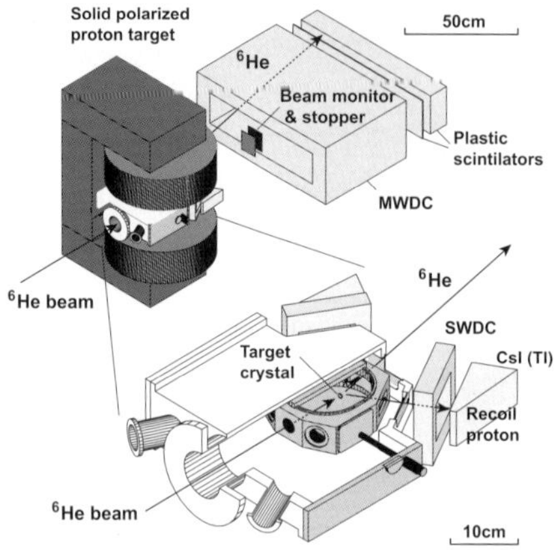

Fig. 1. Schematic view of the detector setup and a polarized proton solid target.

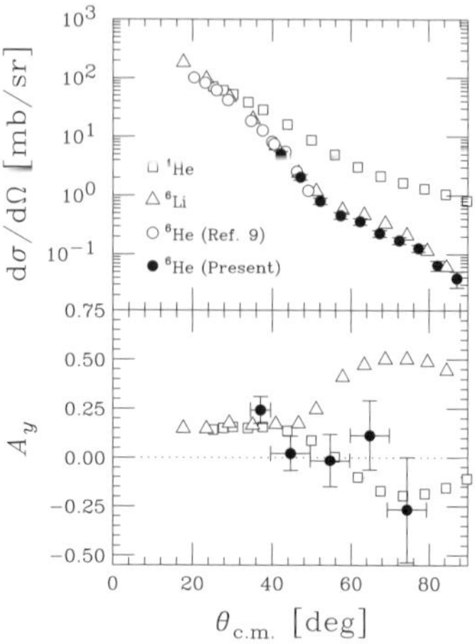

Fig. 2. The $d\sigma/d\Omega$ and A_y of the p+^{6}He elastic scattering at 71 MeV/A (circles), p+^{4}He at 72 MeV/A (squares) in Ref.,[5] and p+^{6}Li at 72 MeV/A (triangles) in Ref.[6]

the gross characteristics of the spin-orbit potential between a proton and ^{6}He, For the central and spin-orbit terms, we assumed the Woods-Saxon type and Thomas type functions, respectively. As an initial potential, a parameter set for p-^{6}Li scattering at 72 MeV/A[6] was used. The dashed and solid lines in Fig. 3 show the calculation with the initial and best- t parameters. First, we discuss the radial dependence of the spin-orbit potential. Figure 4 shows two-dimentional distribution of the radius and di useness parameters of the local (^{6}He,^{12}C,^{16}O) and global (KD, CH89) optical model potentials. The solid contour indicates the simultaneous con dence region of the obtained parameters for p-^{6}He. It is clearly shown that the spin-orbit potential in p-^{6}He is characterized by large radius and/or di useness parameters compared with the global systematics of stable nuclei. Furthermore, the obtained depth parameter, 2.02 MeV, is found to be much smaller than those of global potentials (5 MeV). The phenomenological analysis indicated that the spin-orbit potential between a proton and ^{6}He is characterized by shallow and long-ranged radial dependence. These features of the spin-orbit potential can be naturally explained by the large di useness of the density distribution in neutron-rich ^{6}He.

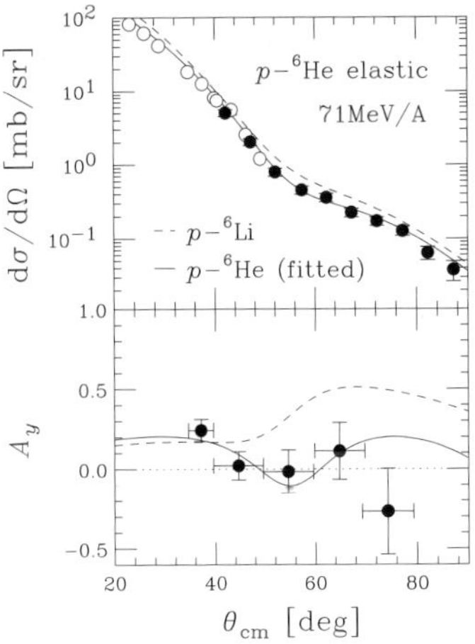

Fig. 3. Results of $d\sigma/d\Omega$ and A_y by the phenomenological optical potentials.

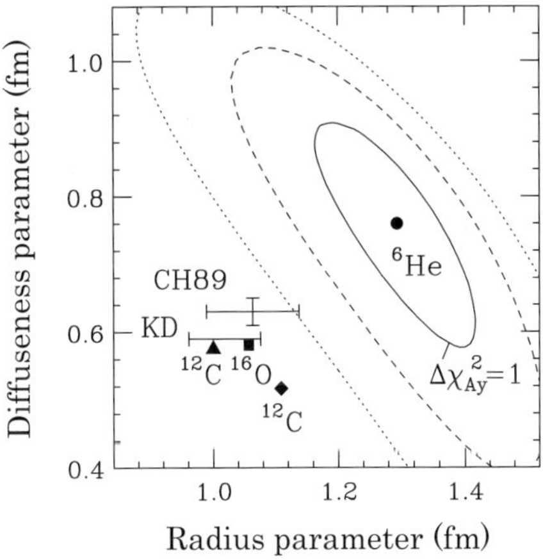

Fig. 4. Two-dimensional distribution of radius and di useness parameters of spin-orbit potentials.

3. Summary and Perspectives

A polarized proton target for RI-beam experiment has been constructed. The vector analyzing powers of the proton elastic scattering from ^{6}He were measured at 71 MeV/A using the polarized target. The spin-orbit potential between proton and ^{6}He is found to be shallower than those in stable nuclei, which is interpreted as an e ect of the di used density in ^{6}He.

It was demonstrated that the analyzing power measurement in RI-beam experiment is e ective in extracting new information on the reaction induced by weakly-bound nuclei. Future studies using polarized proton such as spectroscopy by nucleon knock-out reaction and resonant scattering will enrich our knowledge on the structure of unstable nuclei.

References

1. T. Wakui *et al.*, *Nucl. Instr. Meth. A* **526**, p. 182 (2004); T. Uesaka *et al.*, *Nucl. Instr. Meth. A* **526**, p. 186 (2004); M. Hatano *et al.*, *Eur. Phys. Jour. A Suppl.* **25**, p. 255 (2005).
2. D. Sloop *et al.*, *Journal of Chemical Physics* **75**, p. 3746 (1981).
3. A. Henstra *et al.*, *Phys. Lett. A* **134**, p. 134 (1988).
4. T. Kubo *et al. Nucl. Instr. Meth. B* **70**, p. 309 (1992).
5. S. Burzynski *et al.*, *Phys. Rev. C* **39**, p. 56 (1989).
6. R. Henneck *et al.*, *Nucl. Phys. A* **571**, p. 541 (1994).
7. A. Korsheninnikov *et al.*, *Nucl. Phys. A* **617**, p. 45 (1997).

THE DESIR FACILITY AT SPIRAL2

J.-C. THOMAS
Grand Accélérateur National d'Ions Lourds, CEA/DSM - CNRS/IN2P3, Bd. Becquerel, BP 55027, 14076 CAEN Cedex 05, France

B. BLANK
Centre d'Etudes Nucléaires de Bordeaux Gradignan - Université Bordeaux 1 – UMR 5797 CNRS/IN2P3, Chemin du Solarium, BP 120, 33175 Gradignan, France

The DESIR collaboration proposes the construction of a low-energy beam facility at GANIL-SPIRAL2 to study the properties of exotic nuclei in unexplored regions of the nuclide chart. Beam preparation devices including gas catchers, radiofrequency quadrupoles, high resolution separators and ion traps are under construction in order to provide high quality beams to the users. The DESIR Physics program addresses by means of complementary experimental techniques most of the current interrogations regarding the structure of exotic nuclei, the fundamental interactions driving their properties, as well as their formation in the universe.

Keywords: low-energy facility, nuclear physics, weak interaction, astrophysics

1. Introduction

The DESIR facility is proposed in the frame work of the SPIRAL2 facility presently under construction at GANIL, Caen[1]. SPIRAL2 proposes to its future users a large variety of exotic nuclei produced by fission of ^{238}U, by fusion-evaporation reactions, and by transfer of a few nucleons. The most exotic species – many of which will be produced for the very first time – will be transmitted to the DESIR facility for experiments at low energies that will determine their fundamental, ground-state properties such as decay mode, half-life, mass, charge radius and shape. In addition, DESIR can also receive beams from SPIRAL1, the ISOL radioactive beam facility already existing at GANIL which produces exotic species by projectile and, in the near future, by target fragmentation. Finally, DESIR is also meant to receive beams from the low energy branch of S3[2], the superconducting separator spectrometer proposed for construction in the context of SPIRAL2. A layout of the SPIRAL2 facility is proposed in figure 1.

The DESIR buildings are presented in paragraph 2. Their construction will last from 2012 to mid 2014, when the installation of the beam preparation and experimental equipment can start. The former are described in paragraph 3.

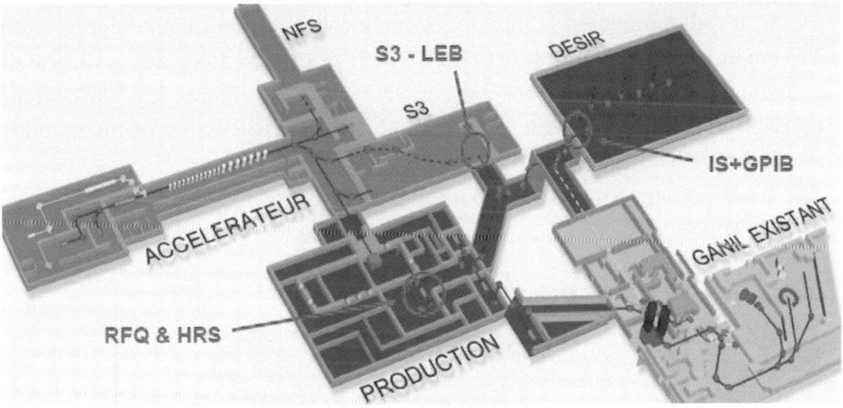

Fig. 1. General view of SPIRAL2 with the DESIR building in the upper right corner and the existing GANIL below. The beam preparation equipment located in the S3 experimental area, in the SPIRAL2 production building and at the entrance of the DESIR experimental room are identified by means of red contours.

They consist (see figure 1) in a radioprotection monitoring station and stable ion sources coupled to a general purpose ion buncher (IS+GPIB). In addition, dedicated beam preparation devices located in the Low Energy Branch part of the S3 facility (S3-LEB) and after the SPIRAL2 production module (RFQ & HRS) will help making the best use of the stable and radioactive ion beams delivered to the DESIR experimental hall.

The experimental equipment comprises three different facilities devoted to the study of the atomic nuclei: The DETRAP facility, which will consist in ion and atomic traps, a laser-spectroscopy facility, LUMIERE and the BESTIOL facility gathering a number of decay spectroscopy devices. The DESIR Physics program is detailed in paragraph 4. The facility is expected to go online by the end of 2015.

2. DESIR buildings

A layout of the DESIR facility is shown in figure 2. The tunnels will allow to transport the beams produced by SPIRAL1, SPIRAL2 and S3 to the DESIR hall where the experimental equipment will be located. The tunnels will have a height and a width of 3m and a total length of 100 m for the three branches.

The DESIR experimental area is a hall of 1500 m^2 with a basement of roughly half this size. The basement will contain the radioprotection and control equipment needed to determine the quality and intensity of the radioactive beams sent into the DESIR hall, a mechanical workshop, clean rooms, assembly rooms, a radioprotection laboratory, storage rooms and different supply rooms. The main experimental area will be located at ground floor and will host the experimental equipment installed at DESIR (see figure 2). The detailed description of the equipment can be found in the DESIR Letter of Intent[3] and in the Technical Design report[4].
The construction of the DESIR buildings should start in 2012 and be delivered by mid 2014.

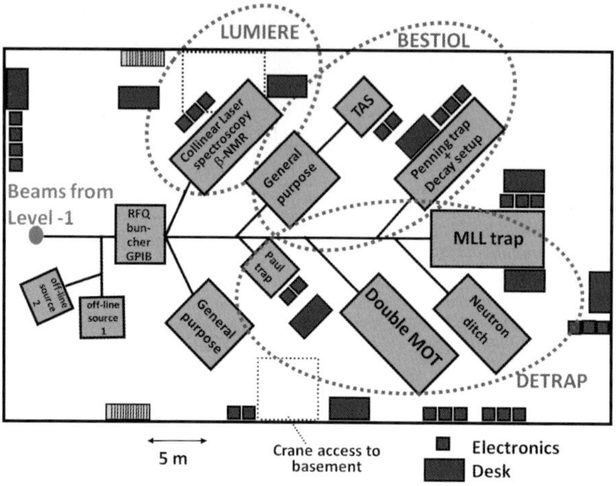

Fig. 2. Schematic view of the DESIR experimental hall (ground level).

3. Beam preparation

In the SPIRAL2 production building (see figure 1), a high-intensity radiofrequency cooler, the SHIRaC device, will be coupled to a high resolution separator (HRS), in order to provide DESIR with high quality mass-separated beams ($m/\delta m \sim 20000$) with energies ranging from 10 to 60 kV and emittances of a few π.mm.mrad.
The low-energy branch of S3 (see figure 1), which will either consist in a radiofrequency gas catcher such as the one designed by the Argonne National Laboratory in the framework of the RIA project, or in an equipment similar to the LISOL device run at Louvain-la-Neuve, Belgium, by the IKS group of

Leuven. In both case, single charged ions, including refractory elements, will be extracted and separated in a HRS similar to the one located in the SPIRAL2 production building before being sent to DESIR.

All the radioactive ion beams sent to DESIR need to be characterized before they are used in the DESIR experimental hall. Such a characterization is also mandatory regarding safety issues. It will be performed at the entrance of the DESIR building by means of an identification station consisting in a fast tape transport station and dedicated charged particle and gamma-ray detectors. The design of the identification unit was performed by the IPHC institute of Strasbourg and a first equipment is currently being used at the ISOLDE facility.

Many experiments need bunched beams in order to increase their signal to noise ratio (e.g. experiments at the LUMIERE facility) or for an efficient injection in their setup (trap-assisted experiments). A general purpose ion buncher (GPIB, see figure 1) will therefore be installed at the entrance of the DESIR experimental hall, in combination with standard stable ion sources.

4. Physics program

4.1. *Beta-decay spectroscopy studies at the BESTIOL facility*

Beta decay offers great insights into the structure of exotic nuclides. In addition, the decay energy, the half-life and the generation of delayed neutrons by fission fragments is of great practical importance for running nuclear power plants. The BESTIOL facility will allow to perform detailed spectroscopy studies of both neutron-deficient and neutron-rich nuclides, with RIB intensities ranging from more than 10^5 pps down to a few ions per hour. It consists in a double Penning-trap isobar separator for trap-assisted spectroscopy in combination with a total absorption spectrometer (TAS), neutron arrays, charged-particle detectors such as the Si Cube and standard beta-gamma detection devices.
The physics cases that will be addressed using the BESTIOL facility are manyfold:

- High-precision measurements of super-allowed and mirror beta decays.
- Lifetime and decay spectroscopy properties of neutron-rich nuclei relevant to nuclear structure and astrophysics studies, as well as to nuclear energy.
- Beta delayed charged-particle correlation studies
- Cluster emission mechanisms

- Shape coexistence, deformation and Gamow-Teller strength around A~100 and at the N=50, 82 closed shells.

4.2. *Collinear Laser spectroscopy at the LUMIERE facility*

Different techniques based on collinear laser spectroscopy as well as experiments based on optically polarized ion beams are proposed using the LUMIERE facility at DESIR. It aims at determining the nuclear moments and spins of nuclear ground-state and long-lived isomeric states, as well as the mean-square charge radius variations over isotopic chains. The sensitivity of the collinear laser spectroscopy technique using optical detection of the fluorescence light is expected to be enhanced by several orders of magnitude using the collinear resonance ionisation spectroscopy method that is currently being developed at ISOLDE.
By using optical pumping inside the cooler, it is possible to access now almost all elements for collinear laser spectroscopy. Pumping in the cooler has been used recently at Jyväskylä to access the Mn isotopes, and can be extended towards and beyond N=40, N=50 and N=82 at DESIR.
By optical pumping using circularly-polarized laser light in the collinear laser beam line, it is possible to reach very large nuclear polarization. Such polarized laser beams can be used for different applications, such as β-NMR studies or for β-γ coincidence spectroscopy to determine asymmetry parameters that are a direct indication for the spin of the populated daughter states.

4.3. *Mass measurement and fundamental interaction studies at the DETRAP facility*

Penning traps, employing highly-homogeneous magnetic and electric fields, now play the leading role for measurements of the nuclear binding energy. Such measurements are crucial to understand the specificities and the evolution of the nuclear structure in the N~Z, A~100 and transactinide regions of the nuclide chart. They are also particularly relevant for the study of the astrophysical rapid-neutron capture process the features of which are intimately related to the mass and decay properties of the isotopes involved. In addition, mass differences allow the precision determination of beta-decay Q values, of capital importance for weak-interaction studies based on super-allowed beta decaying nuclei. Mass measurement will be performed at DESIR with the MLLtrap, which is now being commissioned in Munich. It combines a multi-reflection time-of-flight spectrometer with a Penning trap, leading to a mass resolving power m/δm of about 10^9.

Traps can also be used to hold radioactive species and wait for their decay. The additional property of nicely defining the initial-ion conditions makes traps very effective for the study of decay kinematics and hence correlation coefficients of weak-interaction decays to probe for physics beyond the standard model. The LPCtrap system that has been used for the study of the ^{6}He decay at SPIRAL1 is planned to be installed at DESIR to pursue such studies on selected beta decaying nuclei in order to further probe the contributions of scalar and tensor like currents to the weak interaction. The setup will also be used to measure the beta-neutrino angular correlation coefficient in the beta decay of mirror nuclei.

5. Conclusion

The DESIR collaboration gathers today about 100 scientists belonging to more than 30 institutes worldwide. It proposes to build up a low-energy beam facility at SPIRAL2 where a large variety of high quality radioactive beams, produced by neutron-induced fission, fragmentation, transfer and fusion-evaporation reactions will be available. The complementarity of up-to-date experimental techniques, like laser and trap-assisted spectroscopy, will open unique opportunities for experiments with low-energy exotic nuclei. The construction of the facility should start in 2012 and be operational by the end of 2015.

Acknowledgments

The construction of parts of the experimental equipment was supported by the French ANR via the Contract ANR-06-BLAN-0320.

References

1. http://www.ganil-spiral2.eu/
2. http://pro.ganil-spiral2.eu/spiral2/instrumentation/s3
3. http://www.cenbg.in2p3.fr/desir/IMG/pdf/DESIR LOI LONG.pdf
4. http://www.cenbg.in2p3.fr/desir/IMG/pdf/DESIR-Technical-Proposal-V090105.pdf

ELECTRON SCATTERING

TOSHIMI SUDA
Research Center for Electron-Photon Science, Tohoku University
Mikamine, Sendai, 982-0826, JAPAN

An electron scattering facility dedicated for the structure studies of short-lived nuclei is under construction at RIKEN RI Beam Factory. This facility aims at carrying out never-yet-performed electron scattering experiments for short-lived nuclei to study their electromagnetic properties of the ground state, such as the charge density distributions. The facility is expected to start its operation in the year of 2014.

1. Elastic Electron Scattering

Electron scattering provides the most reliable structure information of atomic nuclei by virtue of the fact that the electron is a point particle, and probes a target nucleus through the fairly weak and well-understood electromagnetic interaction [1]. Therefore, it has been consistently playing a key role in the nuclear structure studies. For example, the best knowledge of the ground-state density distributions of stable nuclei is obtained from a series of elastic electron scattering [2].

It has been, however, strictly limited to stable nuclei (and a few unstable nuclei having very long lifetime), and no experiments for highly unstable (short-lived) nuclei has been ever conducted. This is simply due to obvious difficulties of target preparation for electron scattering.

In view of essential roles played in the structure studies for stable nuclei, electron scattering for short-lived nuclei will undoubtedly provide indispensable information of their internal structure, which is one of the key issues of the modern nuclear physics.

Once electron scattering becomes feasible for short-lived nuclei, the measurements of elastic-scattering cross section will be firstly performed, since its cross section is the largest at low momentum transfer region and the charge density distribution, one of the fundamental ground state properties of nuclei, is determined.

The differential cross section from a spin-less nucleus under PWIA is given as

$$\frac{d\sigma}{d\Omega} = \sigma_{Mott} |F_c(q)|^2, \tag{1}$$

where σ_{Mott} is the Mott cross section, q the momentum transfer and $F_c(q)$ the form factor. The form factor is a Fourier component of the charge density distribution. For further discussion, it is worth noting here that the elastic cross section decreases as $1/q^4$.

For the most stable nuclei, a complete set of their form factors measured up to high momentum transfer is now available [2].

For rarely-produced short-lived nuclei, the cross section measurements will be limited to lower momentum transfer region, since one can not expect that high luminosities are realized. Such measurement will reveal only gross features of the charge distribution, such as only the radius and surface diffuseness. Despite this limitation, however, those radial properties along the isotopic chains would be certainly very important, and be essential *as inputs* for any nuclear structure models applicable for short-lived nuclei.

Based on numerical simulations for elastic scattering off ^{132}Sn which is our first target for electron scattering, it is found that the luminosity of 10^{27} /cm^2/s is required to determine the size and surface diffuseness with the accuracy better than a few %[3].

2. A Novel Technique for Electron Scattering Experiments

2.1. *SCRIT*

To realize the structure studies of short-lived nuclei by electron scattering, we have proposed a novel experimental technique, named SCRIT (**S**elf **C**onfining **RI T**arget) [4], which uses "ion trapping" notoriously known at electron ring facilities. SCRIT aims at forming a target of ions (exotic nucleus) of interest on the electron beam in an electron storage ring.

Ions produced at an external ion source will be transferred to an electron ring, followed by the trap on circulating electron beam by means of the "ion trapping". Since the ions keep trapped transversely on the electron beam, electron scattering off the trapped ions takes place automatically. According to numerical simulation, rather high collision luminosity is expected with small number of trapped ions [4].

To confine the trapped ions also longitudinally, electrodes are placed along the electron beam to form a mirror potential. Controlling the mirror potential, one can inject the target ions from the external ion source, trap them for a certain period depending on lifetime of the target nuclei, and extract them from the trap region for the next injection to keep purity of the target.

2.2. *Feasibility Studies*

In order to demonstrate feasibilities of this novel SCRIT frame, a prototype is constructed and installed at an existing electron ring, KSR, of Kyoto University. Stable Cs ions are used instead of exotic nuclei for this study. Figure 1 shows the SCRIT prototype installed at the 2-m straight line of KSR.

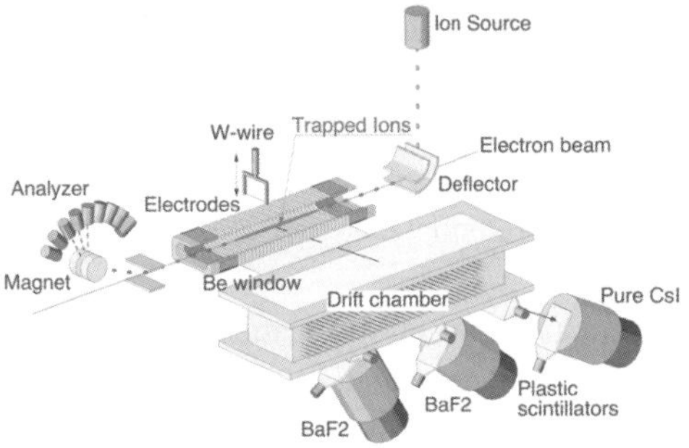

Figure 1. Setup of the SCRIT prototype installed at KSR. It consists of an ion source, a deflector electrode, electrodes for forming the mirror potential, and analyzer and electron detectors.

Cs ions extracted from the ion source are transported and merged with electron beam at the deflector. They are transversely trapped by electron beam, and are longitudinally confined by a mirror potential created by electrodes. By adjusting the mirror potential shape the ion-trapping length was set to 260 mm. The scattered electrons emerging to the air through a thin Be window are detected by an electron detection system consisting of a drift chamber for scattering angle measurement and calorimeters for energy measurement.

We measured scattered electrons from the trapped ions with the ion injection-trap-release cycle by controlling the ion source and the mirror potential. The energy, stored current and lifetime of the electron beam are 120 MeV, 70 mA and 100 sec, respectively. In order to simulate the measurement for short-lived nuclei, the trapping time was set to be 50 msec. Clear signals of elastically scattered electrons from the trapped Cs are observed [5]. The angular

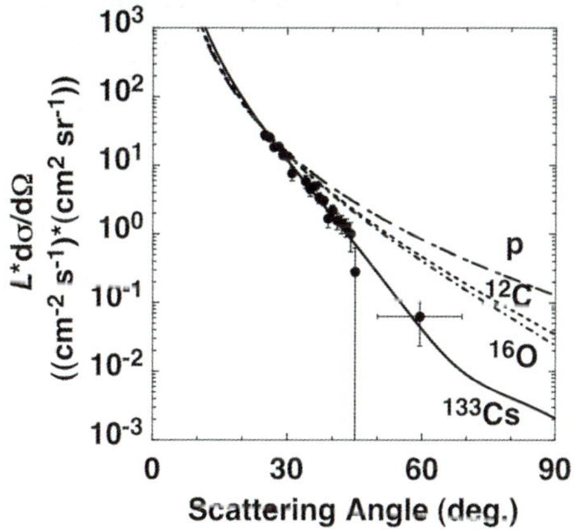

Figure 2. Angular distribution of electron elastic scattering [6]. Solid circles show the elastic events, and the solid line the results of a DWBA calculation for ^{133}Cs. Dashed, dotted, and dot-dashed lines show the elastic cross section for protons, carbon and oxygen, as residual gas components, normalized to the data at a scattering angle of 25° to compare their angular dependence with the experimental data.

distribution, Fig.2, clearly shows that the measurement of elastic scattering using SCRIT is successful [6]. Detailed analysis shows that the luminosity of $1\text{x}10^{26}/\text{cm}^2/\text{s}$ is achieved with 10^6 trapped ions. Since much improved luminosities are expected with larger stored electron-beam current and improved ion-injection technique, we conclude that the luminosity of $1\text{x}10^{27}/\text{cm}^2/\text{s}$ is feasible, and the SCRIT technique is a way to realize never-yet-performed electron scattering of short-lived nuclei.

3. Electron Scattering Facility at RIKEN RI Beam Factory

Based on the success of feasibility studies described above, we started to construct an electron scattering facility at RI Beam Factory. This has been, fortunately, boosted up with the following three "lucky" events;

- Donation of a 700-MeV electron storage ring system from SHI (Sumitomo Heavy Industries), who terminated its use as a light source facility

- FY2008 supplementary budget from the government that enabled us to install the ring at the RIBF
- Grant-in-Aid for Scientific Research (S) has been approved in FY2010 for constructing the an electron detection system

Figure 3 shows a layout of the electron scattering facility at RIBF. The facility consists of an electron accelerator, an ISOL system and an electron detection system.

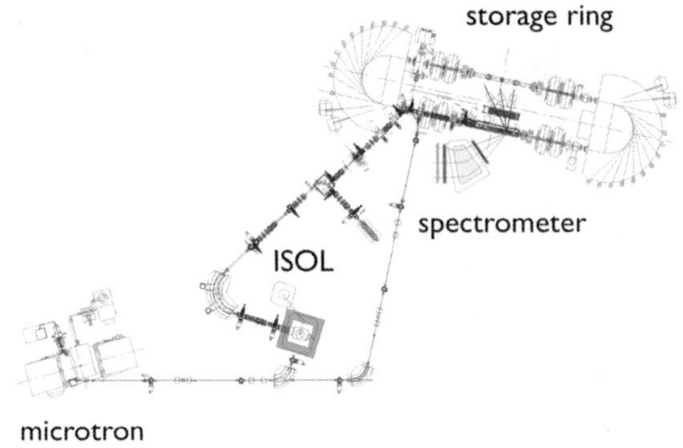

Figure 3. A layout of the electron scattering facility at RIBF partly completed. It consists of an electron accelerator, an ISOL and an electron detection system

3.1. *Electron Accelerator*

The system consists of a 150-MeV injector microtron, the beam transport line and 700-MeV electron storage ring. It was already commissioned, and we are now studying achievable luminosity using stable ^{133}Cs ions at the electron energy of 150 MeV and the stored current of 200 mA.

3.2. *ISOL*

Production of short-lived nuclei, specifically Sn isotopes including ^{132}Sn as the first targets, will be carried out using 150 MeV electron beam bombarding on an uranium carbide target, $^{238}UC_x$. An ISOL system is now under construction. A long beam lifetime (a few hours) of the storage ring will allow us to operate the injector microtron mostly for isotope production.

Using the ^{132}Sn yield ratio in photo-fission process being 1% [7], the production rate of ^{132}Sn is expected to be about 10^8/s at 100-W electron beam power. An efficient extraction with a good separation of the target isotopes, and their pulsing for the SCRIT injection after accumulation are both important R&D issues for coming years.

3.3. *Electron Spectrometer*

In order to identify elastic scattering, an electron spectrometer must have an energy resolution at least better than 1 MeV to resolve elastic and inelastic scattering. The electron beam energy will be 150-300 MeV from considerations of momentum transfer range required for the measurement of the charge density distribution. The momentum resolution, thus, must be better than 3 x 10^{-3}. In addition, a large solid angle must be realized to deal with the small yield from low luminosity experiments.

Due to the fact that the SCRIT provides spatially extended target, typically 40 cm long, any focusing-type high-resolution magnetic spectrometer can not be employed. A non-focusing magnetic spectrometer with tracking detectors as shown in figure 3 will be constructed. This spectrometer covers a wide scattering angular range of 30-60 deg., and has the solid angle of an order of 100 msr.

4. Summary

An extremely unique electron-scattering facility for short-lived nuclei is currently under construction based on a new experimental technique, SCRIT. The first collision between electrons and short-lived nuclei will take place in 2014. The first targets will be Sn isotopes.

References

1. T. deForest, Jr. and J. D. Walecka, *Adv. in Phys.* **5,** 1 (1966).
2. H. de Vries, C. W. deJager and C. deVries, *At. Data Nucl. Data Tables* **36**, 495 (1987).
3. T. Suda and M. Wakasugi, *Prog. Part. Nucl. Phys.* **55**, 417 (2005).
4. M. Wakasugi, T. Suda and Y. Yano, *Nucl. Instrum. Meth.* **A532,** 216 (2004).
5. M. Wakasugi *et al.*, *Phys. Rev. Lett.* **100,** 164801 (2008).
6. T. Suda *et al.*, *Phys. Rev. Lett.* **102,** 102501 (2009).
7. D. De Frenne et al., *Phys. Rev.* **C29**,1908 (1984)

Pushing the limits of Spectroscopy with S^3

B. JP. Gall[1], O. Dorvaux[1], K. Hauschild[2,3], A. Khouaja[1], M. Lamberti[1],
A. Lopez-Martens[2,3], R. L. Lozeva[1], J. Pancin[4] and J. Piot[1], for the S^3 collaboration.

[1] *Institut Pluridisciplinaire Hubert Curien (IPHC), UMR 7178, Université de Strasbourg / IN2P3-CNRS, 23 rue du Loess, 67037 Strasbourg, France*

[2] *Physics Department, University of Jyväskylä, PO Box 35, 40014 Jyväskylä, Finland*

[3] *CSNSM, Université de Paris sud / IN2P3-CNRS, 91405 Orsay Campus, France*

[4] *GANIL, Boulevard Henri Becquerel, 14000 Caen, France*

The new SPIRAL2 LINAG accelerator, developed for production of high intensity deuteron beam, will deliver unprecedented high intensity stable beams. Used for direct reactions at the Super Separator Spectrometer (S^3) target they will open new horizons for the physics of nuclei with low production cross-section and rare nuclei at the extreme limits of the nuclear chart. This article presents the research and development program launched in order to design the best setup according to the three main physics cases of this project: Synthesis of new Super Heavy Elements, Spectroscopy of Very and Super Heavy Elements and Spectroscopy of neutron-deficient nuclei around ^{100}Sn.

Keywords: Spectrometers and spectroscopic techniques, Charged-particle spectroscopy, Windowless silicon detectors, Isomer decay, α-decay.

1. Introduction

SPIRAL2[1] is a large scale international facility dedicated to the production of intense beams of radioactive species by means of fission induced by deuterons in a uranium-carbide target. A superconducting linear accelerator (LINAG), using a latest generation A-PHENIX ECR ion source will provide deuteron beam. In addition, stable beams with intensities as high as 100 pμA for ions with $A < 40 - 50$ will be available.

The Super Separator Spectrometer (S^3) project[2,3] will be one of the first to benefit from these unprecedented intensities of stable beams. It will consist of a high acceptance momentum achromat followed by a mass spectrometer (mass achromat) with a mass resolution approaching 1/400 and a 10^{13} beam rejection power. It opens up a unique opportunity to access the synthesis and spectroscopy of the heaviest elements and the spectroscopy of neutron-deficient nuclei around ^{100}Sn.

Intermediate target surrounded by γ-ray detector arrays such as AGATA[4] , EXOGAM[5] or PARIS[6] could be inserted at focal plane of the S^3 mass achromat for deep inelastic reactions, Coulomb excitation and reaction mechanisms studies.

The full power of the S^3 instrument will be revealed at the final focal-plane of S^3 where a specific and fully optimized identification device will be installed for nuclear structure studies. Alternatively, it will be possible to select recoils with a Gas-Catcher and send them to dedicated physics area such as the DESIR[7] facility.

2. Pushing the limits with S^3

Prompt spectroscopy limit was recently pushed down to the 10 nb scale by the study of the ^{246}Fm nucleus[8] at the University of Jyväskylä using the JUROGAM2, RITU and GREAT[9,10] association equipped with TNT2D[11] digital electronics and a rotating target system. Complementary focal-plane decay spectroscopy studies are limited to the nanobarn scale with the existing setups for fusion-evaporation cross-section of heavy elements.

The LINAG will increase present beams intensities by one to two orders of magnitude. In addition, keeping the rejection power as high as possible, S^3 is optimized for high transmission of several recoils charge states up to final focal-plane. Even if some beam/target combinations may not sustain such high intensities, S^3 will allow physics far beyond present limits.

2.1. *The focal-plane detectors*

Decay studies imply recoiling nuclei, α or β particles, fission, γ-rays or conversion electrons. Thus, powerful selection of rare nuclei of interest is only possible through a combination of several optimized detection devices at the focal-plane of a separator.

Reaction products transported and separated through S^3 will first pass through a time-of-flight (ToF) device in front of, or at, the focal plane depending on the configuration. It will consist of 1 to 2 ToF elements based on Secondary Electron Detectors[12] (SED).

Recoiling nuclei will then be stopped in the implantation Double Sided Silicon Strip Detector (DSSSD). A "tunnel" built with four large silicon detectors placed up-stream of the DSSSD (see figure 1) will detect charged

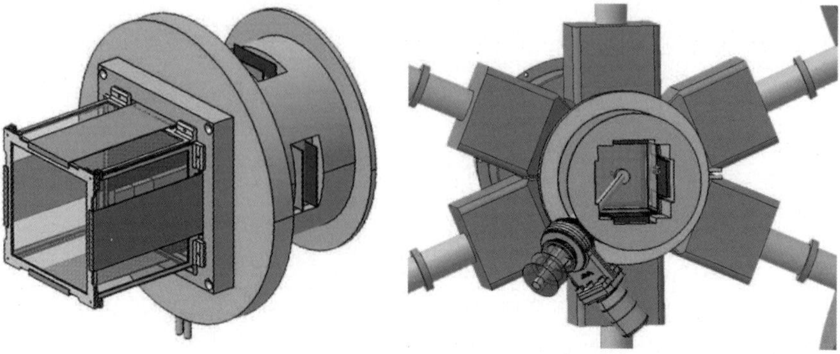

Fig. 1. Optimal arrangement of silicon detectors proposed for the phase 1 setup (left); 3D of the S^3 focal-plane geometry with calibration source in front of the tunnel (right).

particles (α-particles, electrons, protons and fission fragments) emitted during the decay process of the implanted recoiling nuclei. Coincidences between ToF detectors and implantation DSSSD will enable to select the recoiling nuclei and to discriminate between implantation and decay events. A thin "Veto" silicon detector downstream the implantation DSSSD can be used to veto signals from light particles passing through the DSSSD.

The silicon detector array will be surrounded by a germanium array constituted by seven close-packed EXOGAM[5] Clover detectors.

The proposed geometry and relative positioning of the elements as foreseen for the first phase of the project can be seen on figure 1. A wider focal plane, better suited for heavy elements synthesis studies or physics around ^{100}Sn, may be implemented in a forecoming phase of the project.

2.2. *Recoil Tracking and Time-of-Flight measurement*

The different physics cases set strong constraints and limitations on the design and operation of ToF tracking detectors. Trajectory reconstruction can be done through the association of the ToF elements for fast recoiling nuclei where angular straggling is not a key issue. For slow heavy elements produced with actinide targets and light beams it is necessary to use only one ToF element with the thinnest possible emissive foil since straggling and energy loss may be a key issue for their recoil detection.

Time resolution is important for the determination of the A/Q ratio. Timing an position resolution have been tested in-beam with a low-pressure micromegas based ToF prototype and also 1 or 2 dimension SED[12] based ToFs. The 2D SED solution seems to be the most promising.

2.3. *Silicon-detectors array*

The S^3 implantation detector will benefit from feedback from the MUSETT[13] collaboration. It will be based on a "windowless" update of the 10 x 10 cm^2 MUSETT detector with 128 strips per side, since the latest 6" silicon technology enables the production of large detectors with entry window thinner than 50 nm. It allows to work with slow recoiling nuclei, which implant shallowly into the DSSSD.

Optimal length of the silicon tunnel was studied via a GEANT4[14] simulation[15] . Figure 2 shows that very little is gained with a tunnel longer than 10 cm. A 10 x 10 cm^2 tunnel detector based on 0,7 mm thick windowless high resistivity silicon detector will be used.

Better energy reconstruction can be expected for DSSSD-Tunnel coincidences due to the reduction of energy losses in dead layers of the silicon detectors. These effects on the resolution for escaping α particles reconstructed energy were simulated[16] with GEANT4 (see figure 2 middle). Higher granularity is needed in the 2 first centimetres of the tunnel detector where the major part of the α escaped from DSSSD are detected.

A "stripy-pads"[3] prototype single sided windowless silicon detector with 64 equivalent pads (see figure 2 right) is foreseen in order to determine the best capacitance versus optimal granularity compromise for the tunnel. The first prototypes of silicon detectors will be delivered and tested in 2011 and full commissioning is foreseen to start in 2014 at GANIL.

For better cooling, these detectors will be mounted on a ceramic support. Optimal silicon detector operation temperature was determined[17] and optimization of cooling process is ongoing.

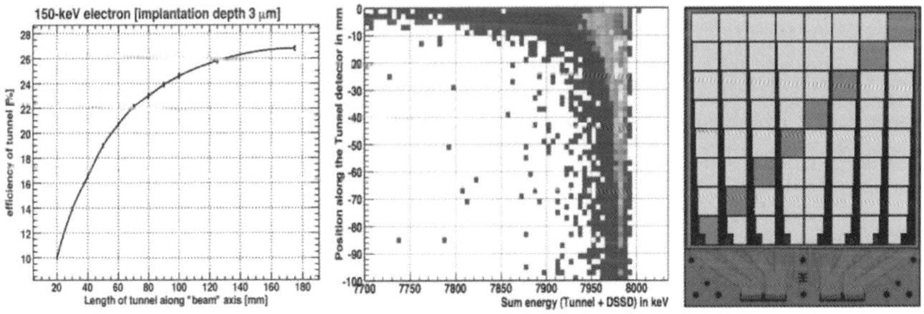

Fig. 2. Simulation of optimal tunnel detectors length[15] (left). Influence with respect to the tunnel hit position of the dead layers of the silicon detectors on a 8 MeV α escaping implantation DSSSD[16] (middle). Tunnel "Stripy-pads" detector with 64 equivalent pixels all read from same side (right); the DSSSD is situated on the top edge of the figure.

2.4. *Focal-plane detectors electronics*

The S^3 project aims at an energy resolution approaching 15 keV for 8 MeV α-particles. In addition, the project needs ablility to distinguish in the tunnel silicon detectors α-particles from conversion electrons and to access to an alpha emitted right after a implantation signal. Triggerless operation with the shortest possible dead time is necessary.

Important efforts are set to develop fast response optimized new electronics such as compensed preamplifier, double gain preamplifiers, integrated solutions based on ASIC's, fast discrete charge preamps...

New digital back-end electronics (BEE) is foreseen to distinguish rare events from transfer reaction products, to trig on fast decay following implantation and to get the ability to record traces for pulse shape analysis.

The physics around ^{100}Sn induces much higher ion rates than SHE synthesis and spectroscopy. Therefore, it implies a higher granularity of the implantation detector resulting in a larger number of electronics channels. It will involve the use of a specific digital electronics based solution.

2.5. *Gamma-ray detection*

The optimization of possible γ-ray detection geometries with existing and/or new dedicated detectors was done[3,18,19] with the GEANT4 code, taking into account realistic recoil-implantation distributions coming from the S^3 optics. A solution based on dedicated new crystals can give up to 63% efficiency for low energy γ-rays. Gamma-arrays based on EXOGAM[5] detectors, one in the back of the DSSSD associated to a ring[3] of 4, 6 or 8 Clovers were studied. The "1+6" Clovers geometry found to be the best solution[3,19] with a 54% total efficiency for low energy γ-rays is quite high as compared to the dedicated germanium solution. This represents more than a factor of 2 with respect to existing focal-planes such as GREAT@RITU[10] and GABRIELA@VASSILISSA[20] . Since the germanium detector directly behind the implantation silicon detector brings half of the efficiency, a dedicated planar detector solution may be proposed in a next phase.

Realistic 3D geometries including all the elements that generate absorption or scattering are in process in order to get realistic simulations of "Day 1" experiments and guide our mechanical and technical choices.

3. Conclusion and perspectives

The Super Separator Spectrometer, recently founded by the French research agency (ANR) as an excellence equipment (EQUIPEX), will make the best use of the unprecedently-high intensity stable beams available from the SPIRAL2 injector. R&D for an optimized focal-plane detection system based on time-of-flight detectors, implantation DSSSD, tunnel silicon detectors, veto silicon, associated electronics and γ-ray array is ongoing. This device will open new horizons for the physics of rare nuclei with low production cross-sections at the extreme limits of the nuclear chart.

4. Acknowledgment

The authors emphasize that S^3 is the work of a wide collaboration from which they are only representatives. They thank all the present and future contributors of this project. A special thanks to E. Gamelin (IPHC) for the 3D conception of the focal-plane elements.

References

1. *Spiral2 White Book Ganil (2006)*, http://pro.ganil-spiral2.eu/spiral2
2. A. Drouart *et al. Nucl. Phys.* **A834**, 747c-750c (2010).
3. B. JP. Gall *et al.*, Acta Phys. Pol. **B42**, 0597-0604 (2011).
4. J. Simpson, J. Phys.: Conf. Ser. **41**, 72-80 (2006).
5. J. Simpson *et al.*, APH N.S., Heavy Ion Physics **11**, 159-188 (2000).
6. A. Maj *et al.*, *SPIRAL2 Letter of Intent: "High-energy γ-rays as a probe of hot nuclei and reaction mechanisms" (2006)*, http://paris.ifj.edu.pl/
7. B. Blank *et al.*, *SPIRAL2 Letter of Intent: "The DESIR facility (Decay, Excitation and Storage of Radioactive Ions)"*, www.cenbg.in2p3.fr/desir
8. J. Piot, PhD Thesis, Strasbourg University 2010 UdS-760;
   J. Piot *et al.*, to be published.
9. P.T. Greenlees *et al.*, Nucl. Phys. **A787**, 507-515 (2007).
10. R. D. Page *et al.*, Nucl. Instrum. Methods Phys. Res. B **204**, 634-637 (2003).
11. L. Arnold *et al.*, IEEE Trans. Nucl. Sci., **53(3)**, 723 (2006).
12. A. Drouart *et al.*, Nucl. Instrum. Methods Phys. Res. A **579**, 1090-1095 (2007);
    J. Pancin *et al.*, J. Inst. **4**, P12012 (2009).
13. F. Jeanneau *et al.*, NIM in preparation,
14. S. Agostinelli *et al.*, Nucl. Instrum. Methods Phys. Res. A **506**, 250-303 (2003); J. Alison *et al.*, *IEEE Trans. Nucl. Sci.* **53**, 270-278 (2006).
15. K. Hauschild, Report to S^3 Detector Working Group in2p3-00431873; http://hal.in2p3.fr/in2p3-00431873/fr/
16. A. Lopez-Martens, Report to S^3 Detector Working Group.
17. R. L. Lozeva *et al.*, NSS-MIC IEEE Symposium Proceedings, 30.10-6.11.2010, Knoxville, USA, in press.
18. M. Lamberti, M1 Research training course, Strasbourg University.
19. A. Khouaja *et al.*, Poster for SPIRAL2 Week 2008 and Collaboration Report.
20. K. Hauschild *et al.*, Nucl. Instrum. Methods Phys. Res. A **560**, 388-394 (2006).

SAMURAI
A Large-Acceptance Spectrometer in RIBF

K. YONEDA

RIKEN Nishina Center,
2-1, Hirosawa, Wako, Saitama 351-0198, Japan
E-mail: kyoneda@riken.jp

SAMURAI is a spectrometer which is now being constructed at RIKEN RIBF. This spectrometer is characterized by a large angular- and momentum-acceptance enabling, for example, multi-particle coincidence measurements. The on-site construction started in October 2010, and the rst experiments will be performed in early 2012. Here the current status and future plan of this SAMURAI project is presented.

1. Outline of SAMURAI Spectrometer

SAMURAI is a spectrometer which is now under construction in RIKEN RIBF.[1] SAMURAI is an abbreviation of "Superconducting Analyzer for MUlti-particles from RAdio Isotope beams with 7 Tm of bending power", and is characterized by a large momentum and angular acceptance, which facilitates multiparticle detection in coincidence.

The main component of SAMURAI is a large superconducting dipole magnet, which has a large gap (80 cm) between the two poles, in which up to 3 T of magnetic eld is applied. This magnet is used for the analysis of projectile-rapidity charged particles which range from protons to projectile-like fragments up to A 100. The large gap provides a large acceptance for particles emitted at forward angles, especially for light particles such as neutrons and protons. Hence SAMURAI is suitable for kinematically complete measurements required in, for example, invariant mass spectroscopy for breakup reactions of loosely-bound nuclei.

The magnet is surrounded by a variety of particle detectors which detect heavy ions, neutrons, and light charged particles such as protons. The detector con guration can be changed depending on experimental requirements. The magnet is rotatable to allow exibile detector con guration.

The magnet construcion will be completed in June 2011. The detectors are being built in parallel. We will get ready for the rst series of experiments, HI-neutron coincidence measurement, in March 2012. In the following, the status of the magnet and some physics topics to be realized with SAMURAI are given.

2. Superconducting Dipole Magnet

2.1. *Magnet Design*

The magnet was designed to ful ll several requircments including:

- Large eld integral for good momentum resolution for projectile-rapidity charged particles
- Large open space between the pole gap for large acceptance
- Small fringe eld for detectors around the magnet

The designed value of the maximum eld integral was set 7 Tm. This corresponds to a rigidily resolution of about 1/700 (rms) for 2.3 GeV/c A/Z 3 particles, which is reasonably good for particle identi cation up to A 100. The pole gap 80 cm corresponds to the vartical angular acceptance of 5 degrees. We adopted the H-type design with cylindorical poles of 2 m diameter. The round-shape superdonducting coils around the poles, each with magnetomotive force of about 1.9 MAT at maximum, provides up to 3 T of magnetic eld at the center of the magnet. The total size of the magnet including the upper, lower, and side yokes is $6.8\ \mathrm{m}^W$ $4.6\ \mathrm{m}^H$ $3\ \mathrm{m}^D$, and the weight amounts to about 700 ton. The eld cramps are attached outside of the coils to reduce the fringe eld less than 50 gauss at 50 cm away from the magnet.

2.2. *Built-in Vacuum Chamber*

The magnet has a built-in vacuum chamber between the gap, so that the particles after the reaction pass through vacuum. The exit window of about 2.4 m 0.8 m has to be covered with a membrane. In order to design the membrane, endurance tests have been made with several types of membranes. A nal choice of window has to be found out, which is strong enough to hold up against vacuum, but with small amount of material, L/L_R $10^{\ 3}$, to reduce multiple scattering and secondary reaction.

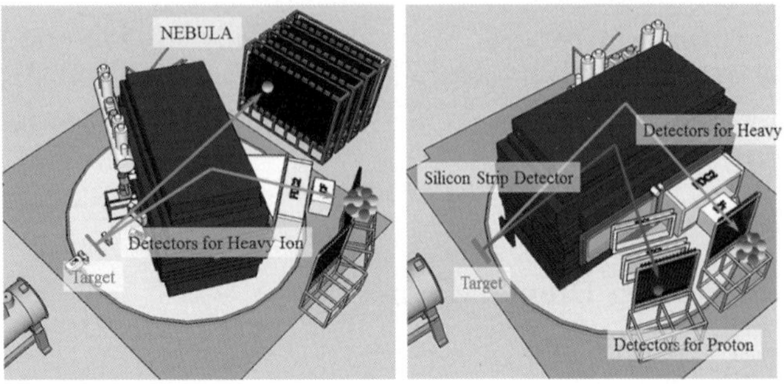

Fig. 1. Typical experimental setup for HI-neutron coincidence measurement (left panel) and HI-proton coincidence measurement (right panel)

3. Experimental Setup

SAMURAI can be used in a variety of pruposes in RI beam experiments. When SAMURAI is used in experiments on the breakup of neutron-rich or proton-rich nuclei, SAMURAI's large acceptance will facilitate e cient HI-neutron and HI-proton coincidence measurements that are required for invariant-mass spectroscopy. SAMURAI will also be suitable for missing-mass spectroscopy, in which the measurement of charged particles after the reaction provides information about the decay modes as well as the tagging of the reaction channels. SAMURAI can also be used to scrutinize reactions of few-nucleon scattering systems such as polarized deuteron scattering on protons, in order to understand fundamental nucleon-nucleon interactions, including three-nucleon force e ects. We also plan to install a time projection chamber (TPC) in the large gap of the SAMURAI magnet, which will be used mainly for reaction studies such as the investigation of the density dependence of the asymmetry term in the nuclear equation of state.

In the following, the experimental setups for two cases will be brie y described; HI-neutron coincidence experiments, and HI-proton coincidence experiments (Fig. 1). These two types experiments are supposed to start in an early stage of SAMURAI operation in our current construction plan.

3.1. *HI-Neutron Coincidence Setup*

HI-neutron coincidence measurement is one of the major usages of SAMURAI, since breakup-type experiment is a powerful tool to pin down the nuclear structure of neutron rich nuclei. In order to perform this type of expepriment, a variety of detectors are being built.

In this type of experiment, RI projectiles hit a target, and break up into neutron(s) and HI residue in a certain probability. The RI projectiles are detected and identified by a plastic scintillator for Time-Of-Flight (TOF) measurement, an ionization chamber for energy loss (E) measurement, and are tracked by two drift chambers (DCs). After the reaction target, the trackes of HI residues in the SAMURAI magnet are measured by DCs before and after the magnet. The HI residues are then detected by an ionization chamber for E measurement and a stack of plastic scintillator hodoscope for TOF measurement which are required for identification of the HI residues.

A neutron detector, called NEBULA (NEutron detection system for Breakup of Unstable-nuclei with Large Acceptance), is placed at forward angles to detect neutrons. Typical distance between the target and NEBULA is 7 m. NEBULA is consisted of 4 layers of plastic scintillator walls, which covers an area of 1.8 m vertical and 3.6 m horizontal. The size of scintillator bar is 120 mm 120 mm 1800 mm, and two photomultiplier tubes are attached on both ends. NEBULA covers the neutron scattering angles up to 5 degrees vertical and 10 degrees horizontal, and realizes 100% coverage up to the relative energy 3 MeV and 40% for relative energy 10 MeV. The detection efficiency for 250 MeV neutrons is 66% with full volume, and 40% with half volume, which will be ready at the start of SAMURAI operation.

3.2. *HI-Proton Coincidence Experiment*

HI-proton coincidence measurement is also to be realized in an early stage, so that we can perform breakup experiments of proton rich nuclei, which provide not only nuclear structure information, but also key information relevant of nuclear astrophysics. A RI beam hits a target, the projectiles break up into protons and reaction residues, and the projectiles and HI residues are tracked and identified by the same detectors used for HI-neutron. For protons, another set of drift chambers are placed after the magnet for track information. A plastic scintillator hodoscope provides TOF and E information.

We are developing a detector between the target and magnet. Although particle positions have to be measured here to deduce scattering angles, the detector is hit by both protons and HIs up to Z 50, and has to be operated for both particles. Since a gas detector cannot be used under this condition, We are preparing a stack of silicon strip detectors. The energy deposit is quite different, up to 2500 times, a development of electric circuits

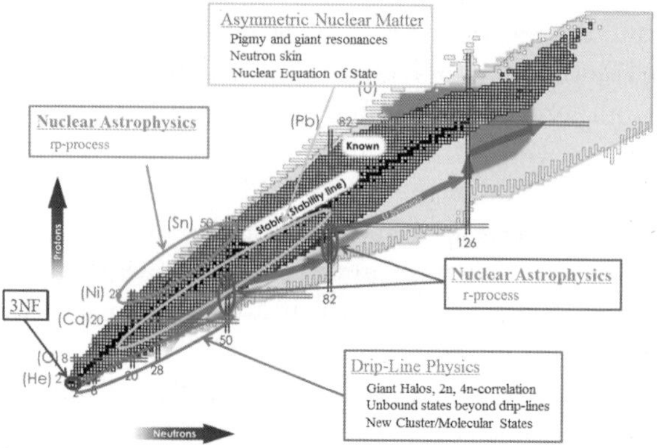

Fig. 2. Physics subjects that SAMURAI will cover in the nuclear chart.

is underway, which covers a dynamic range of 10000 with high density signal processing, by modifying an integrated ASD circuit, called HINP16C, which was developed by a group of Wachington University in St. Louis.

4. Construction Status

A major part of SAMURAI construction was funded over four years beginning from scal year (FY) 2008, and construction of the magnet and detectors is now ongoing. In our current plan, the HI-neutron coincidence measurement will rst be ready in early 2012. After that The HI-proton coincidence will be ready in early 2013, and missing mass measurement, polarized deuteron scattering measuremnt, will be ready in middle 2013. The TPC will be installed in 2014.

5. Physics Topics with SAMURAI

SAMURAI covers a variety of physics subjects related to unbound states of nuclei. Neutron breakup studies can deal with drip-line physics, such as appearance and development of halo structure, 2n correlation in a nucleus, states beyond the drip line, cluster state physics, and so on. Studies of neutron capture rates at N 50 and N 82, which can be deduced through the breakup measurement, will provide important information in understanding the r-process path. In the region closer to the stability, brekup studies can yield information related to asymmetric nuclear matter, such as pigmy and giant resonances, development of neutron skin. The TPC experiments will also provide information on asymmetric nuclear matter, through the studies of nuclear equation-of-state in asymmetric systems. Proton breakup studies will provide key information of the rp-process, which ows in the proton-rich region and plays an important role in studying stellar nucleosynthesis of so-called p-nuclei, which are stable but located in the proton-rich side of the stability line.

These topics also cover a variety of regions in the axis of the excitation energy. In the relatively low excitation energy region, studies of drip-line physics with large neutron-proton asymmetry will be carried out. The pygmy resonance, cluste-state studies correspond to higher excitation energies, and far higher are the giant resonance studies. The studies with the TPC reaches the excitation energy region so high as about 30 MeV

The above physics subjects are all hot topics in the current nuclear physics, and SAMURAI's multiparticle detection capability is well suitable for these unbound state physics studies. SAMURAI will be a useful and fruitful experimental device which yields a lot of physics outcome, and will surely open up new horizon in various aspects of nuclear physics.

6. Summary and Future Outlook

SAMURAI is a spectrometer being constructed in RIKEN RIBF. SAMURAI is characterized by a large angular- and momentum-acceptance, and is suitable for experiments requiring multi-particle detection at forward angles, such as breakup experiments of unstable nuclei.

SAMURAI can be used for a variety of physics subjects by changing experimental setups exibly. HI-neutron coincidence measurements will rst be ready in 2012, which is followed by HI-proton measurement in early 2013. The missing-mass measurement and polarized deuteron measurement will start in middle 2013. A TPC will be installed in 2014.

SAMURAI will surely be a useful experimental device for a variety of RI beam physics. A call for the experiment proposals using SAMURAI will come up soon. It is highly welcomed for scientists who are interested in in SAMURAI-based experiments and developments to join the SAMURAI collaboration, which is currently based on the construction teams for detectors and/or experimental devices, and is going to tramsform into physics collaborations for yielding physics outcome and for building up and enrich physics subjects using SAMURAI.

References

1. Y. Shimizu, J. Phys.: Conf. Ser. **312**, 052022 (2011).

VAMOS: HOT NEWS AND PERSPECTIVES

C. Schmitt* for the VAMOS collaboration

GANIL, CEA/DSM - CNRS/IN2P3, BP 55027, 14076 Caen Cedex 5, France
**E-mail: schmitt@ganil.fr*

The recent upgrade and performances of the VAMOS spectrometer installed at Ganil are presented. The "dual" character of VAMOS combined with its large acceptance makes the device unique worldwide, and of key importance for the efficient use of stable and future radioactive beams at Ganil.

Keywords: Magnetic spectrometer, acceptance, particle identification; LaTeX; Proceedings; World Scientific Publishing.

1. The large acceptance magnetic spectrometer VAMOS

Magnetic spectrometers are currently used for identifying the products of nuclear collisions. A complete characterization of these products in mass M, charge Z and velocity vector v is essential for accurate investigations on nuclear structure and dynamics. Reactions around the Coulomb barrier are very challenging in this respect, requiring a highly efficient and large-acceptance device equipped with an appropriate detection system due to the low energies and wide angular spread of the (weakly populated) ions of interest. The VAriable MOde Spectrometer VAMOS[1] installed at GANIL was originally designed to cope with the challenge inherent to the low-intensity SPIRAL1 radioactive ion beams.[2] High efficiency and isotopic identification of the recoiling ion were among the major goals. Different modes of operation are available depending on the settings of the optical elements. Over the last years, the versatility of VAMOS was further efficiently exploited for the study of various mechanisms such as e.g. deep-inelastic collisions and fission reactions using high-intensity stable beams.
The VAMOS spectrometer basically consists of two quadrupoles (QP) followed by a magnetic dipole (MD). A Wien filter is additionally available if required. The dipole disperses the incoming ions according to magnetic rigidity $B\rho \sim Mv/Q$. The large acceptance of the second QP (100 cm wide along horizontal direction) and the variable deflection angle of the

MD (allowing a variable dispersion at the focal plane) are atypical assets of VAMOS. Also, the whole VAMOS platform can rotate around the target depending on the reaction mechanism of interest. The distance to the target is variable as well, leaving enough space for housing ancillary detectors for prompt e.g. γ or charged-particle spectroscopy. Particle identification in VAMOS is achieved by a combination of detectors located at the focal plane, and giving access to the position (x, y), time-of-flight T_{of}, energy loss ΔE and residual energy E_{Si} of the particle. The effects of the large aberrations inherent to wide-acceptance spectrometers are corrected by software using a sophisticated ray-tracing method, which permits to reconstruct the trajectory of the ion back to the target with high accuracy. The mass and velocity vector of the particle are derived from the measure of the position and time-of-flight, and the software reconstruction. The determination of Z is based on the ΔE-E_{Si} correlation. Note that the measurement of the velocity vector of the particle at the target position is important not only for the study of the reaction mechanism, but it is crucial for nuclear structure investigations based on γ spectroscopy, providing a highly accurate correction of Doppler effects.[3] More details on the particle identification procedure developed at VAMOS can be found in Ref. 1.

2. Prompt γ spectroscopy of isotopically identified fission fragments

As an example of the power of the technique, we focus on a recent experiment[4] dedicated to the isotopic identification and prompt γ spectroscopy of the fragments produced in a fission reaction. The latter was induced in inverse kinematics bombarding a ^{12}C target with a ^{238}U beam at 1.45 GeV. VAMOS was set at 20° with respect to the beam axis to intercept one of the fission fragments. The prompt γ-rays emitted by this fragment as well as by the undetected fission partner were recorded by two EXOGAM clover detectors at backward angles. The fragment entering VAMOS was fully (M, Z, v) identified as outlined above. In addition, the velocity vector of the *undetected* partner was reconstructed using two-body kinematics. The two possibilities for properly Doppler-correcting the γ-ray spectrum are thus available. The chart of Fig.4 (relegated to the end of the manuscript for sake of place) shows the high quality of the identification over the full and wide (M, Z) range spanned by fission. The isotopic distribution of the fragment is displayed as a function of the atomic charge. Note that inverse kinematics is of great help here, increasing the detection efficiency and facilitating the identification in Z (owing to the forward focusing and high kinetic energy of the products, respectively, as compared to direct kinematics). Resolutions of $\Delta M/M \sim 160$ and $\Delta Z/Z \sim 65$ were achieved. The power of the coupling of VAMOS with EXOGAM is illustrated in Fig.1. The left panel presents the γ-ray spectrum measured in coincidence with ^{138}Xe detected in VAMOS and corrected with the corresponding measured velocity of ^{138}Xe. The characteristic γ lines of this particular isotope show up as sharp peaks, while broad peaks are due to transitions coming from the undetected fission partner which energies are wrongly Doppler corrected. Yet, studying the spectroscopy of the undetected fission partner can be done as well with the present technique. This is demonstrated in the right panel of Fig.1 where the γ-ray spectrum measured in coincidence with ^{133}Xe identified in VAMOS and corrected with the reconstructed velocity of the undetected fragment is displayed. The γ-ray lines of the various Ru isotopes, which are the fission partners of ^{133}Xe, are clearly visible. Further details on the power of the technique over other methods (namely $\gamma - \gamma$ methods are reported in Ref. 5).

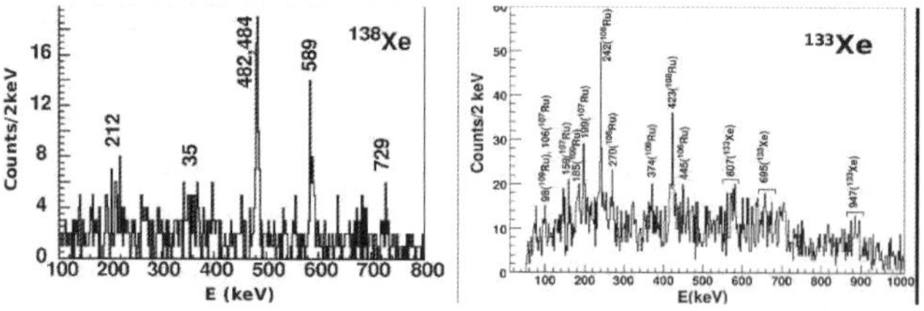

Fig. 1. Left: γ-ray spectrum measured in coincidence with ^{138}Xe identified in VAMOS and corrected with the corresponding measured velocity. Sharp γ lines belong to ^{138}Xe. Right: γ-ray spectrum measured in coincidence with ^{133}Xe identified in VAMOS and corrected with the reconstructed velocity of the undetected fission partner. Figures are adapted from Ref. 5.

3. Recent doubling of the acceptance

Although among the largest available, the acceptance of VAMOS remained till recently limited, and considerably dependent on the momentum of the particle.[1] Accurate investigations based on ion-optical calculations showed that the acceptance was primarily limited by the size of the detectors (and *not* by the performance of the QP's). These calculations further predicted that doubling the size of the focal plane would permit to fully exploit the capabilities of the quadrupoles and make the acceptance uniform, nearly independent on dispersion. A new large-area detection system was therefore built. It comprises: a Multi-Wire Parallel Plate Avalanche Counter (MW-PPAC), two Drift Chambers (DC), a segmented Ionization Chamber (IC) and a wall of 40 Silicon detectors (Si). Each detector has an active area of 100 cm × 15 cm. While the DCs give access to the position (x, y), the IC and Si wall measure the ΔE and E_{Si}, respectively. The stop signal of the T_{of} is provided by the MWPPAC. A smaller (10 cm × 6 cm large) MWPPAC was additionaly inserted between the target and the entrance of the spectrometer. When used as the start for the T_{of}, it permits to improve on the mass resolution (being so far dependent on the timing properties of the beam). Special attention was paid on minimizing the amount of dead layers along the trajectory of the particle, providing accurate particle identification of heavy ions produced in various mechanisms around the Coulomb barrier. The details on the design of the detectors and associated electronics can be found in Ref. 6.

The commissioning experiment of the upgrade detection system used a ^{129}Xe beam at 967 MeV impinging on a gold target. The spectrometer was rotated to 40° with respect to the beam direction. The IC was operated with CF_4 gas at a pressure of 30 mbar to optimize the resolution of the high Z elements produced in deep-inelastic collisions. The Z resolution can be appreciated from the left panel of Fig.2 where the measured ΔE-E_{Si} correlation is shown. A value of $\Delta Z/Z \sim 70$ was extracted. The mass distribution is displayed in the right panel. An improved resolution of $\Delta A/A \sim 220$ was achieved owing to the small MWPPAC close to the target. With the new focal plane detectors, the momentum acceptance was found to be increased from ±10% up to nearly ±20% for a solid angle of 60 msr.[6]

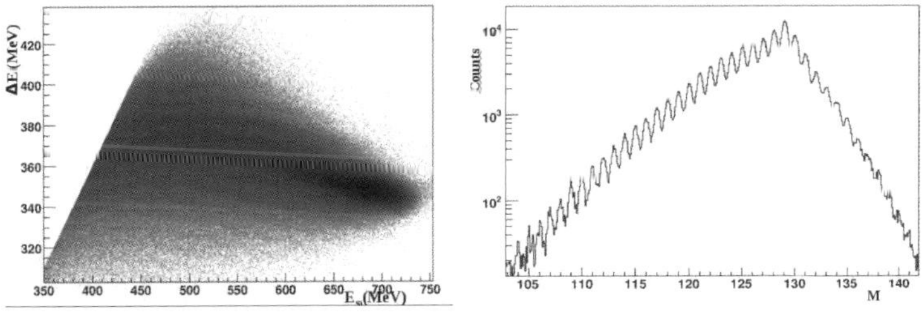

Fig. 2. Left: Correlation ΔE-E_{Si} for the fragments produced in ^{129}Xe +^{197}Au collisions. Right: Corresponding mass distribution. Figures are adapted from Ref. 6.

4. Gas-filled mode of operation

The above discussed new detection system enables exploiting the full capability of the vacuum spectrometer. Yet, a great part of the physics program planed with the future high-intensity SPIRAL2 fission fragments beams at Ganil is based on fusion-evaporation reactions in symmetric and inverse kinematics. For such configurations the use of the vacuum spectrometer is limited, due to the insufficient rejection of the incoming beam at 0°. A pilot experiment was recently performed demonstrating that VAMOS can also act as a powerful separator once filled with a gas at low pressure.[7] According to charge exchange properties in the gas, the particles focused around 0°, i.e. beam and evaporation residues in a fusion reaction, acquire magnetic rigidities which are different enough to be physically separated at the focal plane (contrary to the situation in vacuum).The focal plane detectors are placed such as to intercept efficiently the fusion products, while avoiding the intense incoming beam.

To test whether the concept behind gas-filled magnets can be implemented at a large-acceptance spectrometer such as VAMOS, the aforementioned standard set-up was modified as follows. A carbon foil was placed 1 m before the target to separate the vacuum of the beam line from the gas-filled region. The 8 meter-long VAMOS was filled with Helium at about 1 mbar pressure, and a Tantalum plate was set at the end of the dipole chamber for dumping the rejected beam. A ^{40}Ca beam at 196 MeV was used to bombard a ^{150}Sm target. While the direct beam was efficiently rejected by the gas filling (rejection factor estimated larger than $10^{1}0$), fusion-evaporation residues were further properly discriminated from remaining un-wanted particles using the correlations of various observables measured at the focal plane, namely time-of-flight, energy loss and residual energy. Figure 3 illustrates the clean selection of the ions of interest achieved with the correlation ΔE-E_{Si} as an example. These residues were isotopically identified by their characteristic prompt γ-rays measured by EXOGAM detectors placed around the target, and/or by their characteristric radioactive α-decay after implantation in the Si wall at the focal plane. In addition, the software reconstruction of the trajectory was successfully done, providing the suited event-by-event Doppler correction of the γ-ray spectrum. This was never attempted at any gas-filled separator, and a mean velocity correction was commonly applied. Detailed ion-optical calculations were used to estimate the transmission of the gas-filled VAMOS: Values larger than 80% and 95% were obtained for αxn and $xnyp$ evaporation channels, respectively. These numbers are up to a factor of two as large as what currently achieved at other devices.[7]

The main reason of this difference is to be attributed to the large entrance aperture of VAMOS.

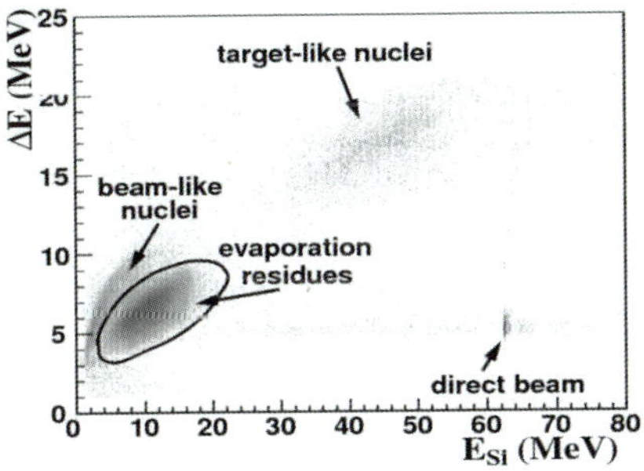

Fig. 3. Correlation ΔE-E_{Si} for the products of the ^{40}Ca +^{150}Sm reaction at the gas-filled VAMOS. The various contributions are indicated. Figure borrowed from Ref. 7

.

5. Conclusion

The performance of the large-acceptance spectrometer VAMOS of Ganil for studies of nuclear structure and reaction mechanism around the Coulomb barrier are presented. The capabilities of the device when coupled to an efficient γ-array for studying exotic nuclei is first highlighted by data on the prompt γ spectroscopy of isotopically identified fission fragments. The recently built new detection system, twice as large as the previous one, is further described. It permits to double the acceptance of the spectrometer and to make it dispersion-independent. The power of the new system is demonstrated with the result of the commissioning experiment dedicated to the detection of deep-inelastic products formed in ^{129}Xe +^{197}Au collisions. Finally, the feasibility of a performant gas-filled mode of operation of VAMOS are demonstrated with the results of a recent test experiment in which the fusion evaporation residues produced in ^{40}Ca +^{150}Sm collisions were successfully separated from un-wanted particles at 0°. The availability of such a highly selective separator opens a window to explore the potential of fusion reactions over a wide dynamical range and in varying kinematical conditions, with unequalled transmission.

6. Acknowledgments

The authors would like to thank the GANIL management and technical staff for their full support of the various projects and smooth running of the corresponding experiments, respectively.

References

1. S. Pullanhiotan et al., Nucl. Inst. Meth. Phys. Res. A 593 (2008) 343.
2. H. Savajols, VAMOS Collaboration, Nucl. Phys. A 654 (1999) 1027c.
3. S.Bhattacharyya et al., Phys.Rev.Lett. 101 (2008) 032501.
4. M. Caamano et al., Proc. 4th Intern. Workshop Nuclear Fission and Fission-Product Spectroscopy, Cadarache, France, 13-16 October 2009, A.Chatillon, H.Faust, G.Fioni, D.Goutte, H.Goutte, Eds., p.15; AIP Conf.Proc. 1175 (2010).
5. A. Shrivastava et al., Phys. Rev. C 80 (2009) 051305(R).
6. M. Rejmund et al., Nucl. Inst. Meth. Phys. Res. A 646 (2011) 184.
7. C. Schmitt et al., Nucl. Inst. Meth. Phys. Res. A 621 (2010) 558.

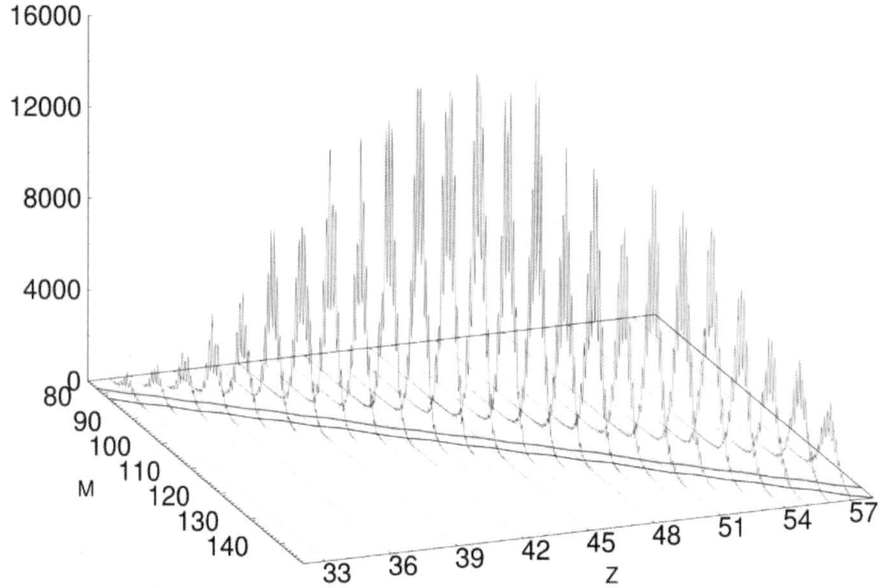

Fig. 4. Mass M and charge Z identification chart of the fission fragments produced in $^{238}U + ^{12}C$ collisions.

Final Remarks

Takaharu Otsuka

Department of Physics, University of Tokyo,
Hongo, Bunkyo-ku, Tokyo 113-0033, Japan
**E-mail. otsuka@phys s.u-tokyo.ac.jp*

Final remarks of this symposium are given. Physics of exotic nuclei with RI-beam is rapidly developing, and many exciting results and ideas were presented in this symposium. I also noted that the observation of the second 0^+ state of ^{34}Si isotope is of particular interest.

Keywords: RI-beam, exotic nuclei

First of all, I found that this symposium has been very successful in various respects. Reflecting that the field of physics with rare isotope beam is extremely active,[1] the symposium has been very timely and exciting with many new results and ideas. I appreciate efforts of all participants for their attendance, particularly those who have come from France over such a distance. In my talk, *i.e.*, final remarks, I have shown forty one slides many of which were copied from original presentations in the symposium. There were so many exciting presentations, and it is clearly impossible or at least useless to summarize all of them in a short article. If I dare to do so in a lengthy article, it would be redundant with actual proceedings reports. So, I am not going to mention them all in this article. It is stated, however, that I was so impressed by recent advances of nuclear physics, which are largely due to GANIL and ISOLDE and also due to RIBF of RIKEN. We look forward to the completion of SPIRAL-2 of GANIL as well as other RI-beam facilities being constructed in the world. The field keeps growing.

I would like to mention several presentations in which I was particularly interested and/or impressed. One of them was the performance of VAMOS with fission fragments. Recent and future progress in RIBF sounds very productive, and should be well collaborated by France and Japan. As an example of joints projects between France and Japan, MINOS looks very promising. The measurements of magnetic moments both in RIBF and

GANIL show the importance and efficiency of international collaboration. In addition, as a personal feeling, I realized that a good fraction of the presentations in this symposium are experimental studies on the shell evolution in exotic nuclei, which is primarily due to the tensor and three-body forces.[2–6]

Among many important and interesting presentations, I was particularly pleased to see the second 0^+ state of ^{34}Si. This state has been unknown for decades, and its excitation energy has been debated by different theoretical calculations. Several "observed" levels had been reported, but have not been confirmed. Now the situation has become clear. The group of Dr. Grevy has succeeded in locating this state by a genius idea. Since there is no proceedings article on this talk, I briefly sketch how the measurement was done without going too much in detail. The β-decay from ^{34}Al is used. For ^{34}Al, an isomer has been predicted actually by my group in terms of the Monte Carlo Shell Model[7] at around 200 keV with $J^{\pi}=1^+$. Since levels below are predicted to be 4^- and 5^-, this 1^+ state cannot decay fast by γ decay, becoming an isomer. The Gamow-Teller β-decay from this state produces a 0^+ state of ^{34}Si. By using this β decay as a trigger, e^+ and e^- from the E0 transition down to the ground state were observed. This gives us the energy of the second 0^+ state. It is about 2700 keV, and is considerably below the 2^+ state.

I just show our prediction[8] in Fig.1. Figure 1 shows both the 2_1^+ and 0_2^+ levels. These levels were predicted somewhat too low but their relative pattern turned out to be rather close to experimental one. We have been worried about the relative location of the 0_2^+ level relative to the 2_1^+ level, as this was the major known open problem with our SDPF-M interaction. I am thus very much pleased with the result, but my appreciation is not because of our prediction but because of the genius experimental approach to this long-standing problem.

I hope that we will get together in the next French-Japanese joint symposium with new results and emerging ideas like this symposium. Last but not least, I would like to express sincere gratitude on behalf of all the participants to *real* organizers of this symposium, Dr. H. Otsu, Dr. T. Motobayashi and Dr. P. Roussel Chomaz.

References

1. A. Gade and T. Glasmacher, Prog. Part. Nucl. Phys. **60**, 161 (2008); O. Sorlin and M.-G. Porquet, Prog. Part. Nucl. Phys. **61**, 602 (2008).
2. T. Otsuka *et al.*, Phys. Rev. Lett. **87**, 082502 (2001).

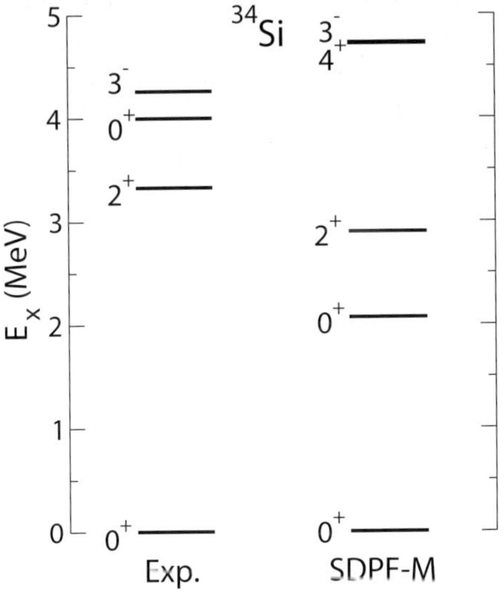

Fig. 1. Energy levels of ^{34}Si. Experimental levels are taken from earlier works, with a wrong second 0^+ state. Theoretical results are from Monte Carlo Shell Model with SDPF-M interaction. The experimental 0_2^+ level is taken from Nummela *et al.*,[9] and contradicts the report by Gravy in this symposium.

3. T. Otsuka *et al.*, Phys. Rev. Lett. **95**, 232502 (2005).
4. T. Otsuka *et al.*, Phys. Rev. Lett. **104**, 012501 (2010).
5. N. Tsunoda *et al.*, Phys. Rev. C **84**, 044322 (2011).
6. T. Otsuka, T. Suzuki, JD.Holt, *et al.*, Phys. Rev. Lett. **105**, 032501 (2010).
7. P. Himpe, G. Neyens, D.L. Balabanski, *et al.*, Phys. Lett. B **658**, 203 (2008).
8. Y. Utsuno, T. Otsuka, T. Mizusaki and M. Honma, Phys. Rev. C **64**, 011301(R) (2001).
9. S. Nummela, *et al.*, Phys. Rev. C **63**, 044316 (2001).